Rasmus Linser

MAS Solid-State NMR on Biomolecules

AF092476

Rasmus Linser

MAS Solid-State NMR on Biomolecules

Development and Application of New Methodology

Südwestdeutscher Verlag für Hochschulschriften

Imprint
Any brand names and product names mentioned in this book are subject to trademark, brand or patent protection and are trademarks or registered trademarks of their respective holders. The use of brand names, product names, common names, trade names, product descriptions etc. even without a particular marking in this work is in no way to be construed to mean that such names may be regarded as unrestricted in respect of trademark and brand protection legislation and could thus be used by anyone.

Publisher:
Südwestdeutscher Verlag für Hochschulschriften
is a trademark of
Dodo Books Indian Ocean Ltd., member of the OmniScriptum S.R.L Publishing group
str. A.Russo 15, of. 61, Chisinau-2068, Republic of Moldova Europe
Printed at: see last page
ISBN: 978-3-8381-2548-0

Zugl. / Approved by: Berlin, HU, Diss., 2007

Copyright © Rasmus Linser
Copyright © 2011 Dodo Books Indian Ocean Ltd., member of the OmniScriptum S.R.L Publishing group

WHAT THIS WORK IS ABOUT

This thesis focuses on the methodological improvement of proton detected solid state NMR on proteins. The understanding of biological processes on an atomic basis is the objective of a great part of current scientific research. The manifold potential of such knowledge ranges from a fundamental understanding to medical and technical applications. In respect to this aim, elucidation of structure and dynamics of involved proteins, membranes or nucleic acids is essential. One respective tool, employed especially for systems that are difficult to crystallize or solubilize, is solid state NMR spectroscopy. Traditionally, solid state NMR suffers from both a low resolution and sensitivity especially for large bio-molecules. A partial proton exchange against deuterons can yield a manifold enhanced resolution. Yet, sensitivities formerly obtained with extensively proton-diluted samples have not allowed for the acquisition of basic NMR experiments for sequential assignment or structural restraints.

The dissertation focuses on conceptual improvements of this methodology. In particular, preparative and spectroscopic techniques are presented that allow high-resolution solid state NMR together with a high sensitivity. An improved sensitivity was provided by Paramagnetic Relaxation Enhancement (PRE), which was optimized for the application to micro-crystalline proteins. This technique, which allowed to speed up data acquisition for proton diluted samples by a factor of up to 15, is described in Chapter 3. Here, the development of a derived method is elucidated which allows to determine the surface accessibility of an amide proton in a micro-crystalline protein. As a second prerequisite for a high sensitivity, the knowledge about the optimal proton content for a specific purpose was obtained. This is described in Chapter 4.1.

Provided a strongly increased sensitivity, new experiments yielding backbone and sidechain assignments based on the employed labelling were developed. These experiments, which play a fundamental role for spectroscopy on deuterated proteins, are described in Chapter 4.2 to 4.4. Here, a particular attention is paid to the resonance assignment of residues undergoing slow motion (Chapter 4.4), which are often important for protein function. In addition to these experiments yielding the complete ^{13}C, ^{15}N, and ^{1}H assignments of a proton diluted protein, an improved methodology for the acquisition of $^{1}H/^{1}H$ distances was developed (Chapter 4.5).

The described experiments were developed using the SH3 domain of chicken α-spectrin, but are applicable to a wide range of targets. In Chapters 5 and 6, application to the 40-residue peptide

What this work is about

$A\beta^{1-40}$, a plaque forming biomolecule occurring in the course of Alzheimer's disease, and to the 280-residue β-barrel protein OmpG is shown.

The developed strategies based on a defined proton dilution in solid state NMR have been shown to be an important step for the applicability of solid state NMR in structural biology already now: The characterization of H-bond topologies in a micro-crystalline protein, e.g., which was recently presented by Schanda et al.,[1] is based on the developments of this thesis.

TABLE OF CONTENTS

What This Work is About 1

Table of Contents 3

Abbreviations 5

1 Introduction 7
    1.1 Cellular processes as determinants of life 7
    1.2 The physics of NMR and its quantum-mechanical description[32-34] 12
    1.3 Relaxation[33] 20
    1.4 Traditional solid state NMR[34] 22
    1.5 Proton chemical shifts in solid state NMR 24

2 Proteins and Preparation 27
    2.1 The SH3 domain of chicken α-spectrin as a system for solid state NMR methods development 27
    2.2 $A\beta^{1-40}$, a fibril forming 40-residue fragment of the Amyloid Precursor Protein APP 30

3 Paramagnetic Relaxation 33
    3.1 Paramagnetic Relaxation Enhancement (PRE) 35
    3.2 Surface accessibility in the solid state 39

4 Methodological improvement of ^{1}H detected solid state NMR 49
    4.1 The proton content 49
    4.2 Solution state like 3D experiments 52
        3D experiments for backbone assignment 53
        Sequential assignment via CO chemical shifts 56
        Sequential assignment via C^{α} and C^{β} chemical shifts 63
        CP-based triple resonance experiments 69
    4.3 Differential relaxation and spin state selection 69
    4.4 Side chain assignment 79
        Full-side chain correlations 80
        Side chain amide correlations 84
    4.5 Magnetization transfer through space – proton RFDR experiments 90

Abbreviations

   Cross peak build-up 93
   Backbone assignment by sequential contacts 97
   Solid state NOESY using scalar correlations 98
  4.6 Application of improved methodology in ^{1}H detection to the SH3 domain of chicken α–spectrin 100
   Reassigned resonances 102
   New assignments 103

5 Proton detected solid state NMR applied to Aβ$^{1-40}$ 111
  5.1 Sample preparation and spectral quality 112
  5.2 Backbone assignment 115
  5.3 Side chain spectroscopy 124
   CIDNP type effects observed for Aβ$^{1-40}$ fibrils at elevated temperatures 125

6 Proton detected solid state NMR applied to membrane proteins 128

7 Discussion and Conclusions 131

References 136

Appendix 146
   Out-and-back HNCO using INEPTs and WALTZ-16 decoupling 146
   Out-and-back HNCO using CPs and WALTZ-16 decoupling 149
   Out-and-back TROSY -HNCO 152
   Out-and-back HNCACB using INEPTs and WALTZ-16 decoupling 155
   Out-and-back HNCACB using CPs and WALTZ-16 decoupling 159
   Out-and-back HNCACO/HNCOCA using INEPTs 162
   Full side chain correlation experiment using H/N-INEPTs 166
   CaCOH experiment for side chain amide assignments 169
   RFDR-HSQC experiment, CP based 172
   HSQC-RFDR-HSQC experiment, CP based 174
   HSQC-RFDR-HSQC experiment, INEPT based 177

Acknowledgements 181

Publications Arising from this Work 183

Contributions to Conferences 184

ABBREVIATIONS

ali.	Aliphatic
APP	Amyloid Precursor Protein
Aβ	amyloid β
BACE	β-site APP cleaving enzyme
CDK	Cyclin Dependent Kinase
CODEX	Centerband-Only Detection of Exchange
CIDNP	Chemically Induced Dynamic Nuclear Polarization
CP	Cross Polarization
DARR	Dipolar Assisted Rotational Resonance
DNA	desoxy ribonucleic acid
DO2A	1,7-dicarboxymethyl-1,4,7,10-tetraazacyclododecane
DREAM	Dipolar Recoupling Enhanced by Amplitude Modulation
DTPA	Diethylene triamine pentaacetate
E. coli	*Escherischia coli*
EDTA	ethylenediaminetetraacetate
FID	Free Induction Decay
FSLG	Frequency-Switched Lee-Goldberg
GPCR	G-Protein Coupled Receptor
H-bond	Hydrogen-bond
HMQC	Heteronuclear Multiple Quantum Correlation
Hsp	Heat Shock Protein
HSQC	Heteronuclear Single Quantum Correlation
HWFM	Half Width at Full Maximum
INEPT	Insensitive Nuclei Enhanced Polarization Transfer
IPTG	Isopropyl-β-D-thiogalactopyranoside
LB	Luria Broth
MAS	Magic Angle Spinning
MISSISSIPPI	Multiple Intense Solvent Suppression Intended for Sensitive Spectroscopic Investigation of Protonated Proteins, Instantly
NADPH	Nicotinamide adenine dinucleotide phosphate-H^+
NMR	Nuclear Magnetic Resonance

Abbreviations

NOE	Nuclear Overhauser Effect
NOESY	Nuclear Overhauser Effect Spectroscopy
OD	optical density
OmpG	Outer Membrane Protein G
PAIN-CP	Proton Assisted Insensitive Nuclei Cross Polarization
PDSD	Proton Driven Spin Diffusion
PMLG	Phase-Modulated Lee-Goldberg
PRE	Paramagnetic Relaxation Enhancement
R^2	Rotational Resonance
RDC	Residual Dipolar Coupling
REDOR	Rotational Echo Double Resonance
RFDR	Radio Frequency Driven Recoupling
RMSD	Root Mean Square Deviation
RNA	ribonucleic acid
sc.	side chain
SH3	Scr-Homology 3
SPECIFIC-CP	SPECtrally Induced Filtering In Combination with Cross Polarization
ST2-PT	Single Transition to Single Transition Polarization Transfer
TALOS	Torsion Angle Likelihood Obtained from Shifts and sequence similarity
TEM	Transmission Electron Microscopy
Tris	*Tris*(hydroxymethyl)-aminomethane
TROSY	Transverse Relaxation Optimized Spectroscopy
VDAC	Voltage Dependent Anion Channel
WALTZ	Wideband, Alternating-phase, Low-power Technique for Zero-residual-splitting

1 INTRODUCTION

1.1 Cellular processes as determinants of life

Although living organisms are built from just the same elements as non-living matter, an amazing number of peculiar processes arrange the complex molecular functionality necessary for a living being to grow, move, adapt and reproduce itself. On the atomic level, all these organisms share a common strategy, using special types of molecules like nucleic acids for the storage and procession of genetic information, proteins for metabolic processes, signal transduction, movement e. g., sugars and phospholipid molecules for construction of structural elements serving as a scaffold for the other processes etc.. In order to allow for an efficient variation and adaptation of existing machinery to new evolutionary requirements, nucleic acids and proteins each consist of simple building blocks.

Proteins are probably the most diverse machinery, they perform various tasks enabling e. g. intra- and intercellular communication processes, they provide all the tools for a transport, chemical modification, decomposition and construction of molecules, and they are in charge of physical action of the cell, like for muscle contraction or temperature preservation. The building blocks of proteins are the amino acids, molecules of around 100 Da, that can be linked together by stable peptide bonds. These bonds provide a certain rigidity due to the involvement of sp^2-hybrid orbitals and additional p-orbitals forming partial C-N double bonds. The different amino acids provide a diverse set of side chains with different chemical properties in terms of charge, polarizability, acidity and aromaticity e. g.. These contribute to a defined overall fold of the protein including reaction centers, binding interfaces or selective passages for small molecules, based on electrostatic, steric, and van-der-Waals interaction. Figure 1.1 displays glutamine as a neutral amino acid with manifold function involved in e. g. solubility of proteins, salt bridge formation, catalysis processes including acid/base reaction etc.

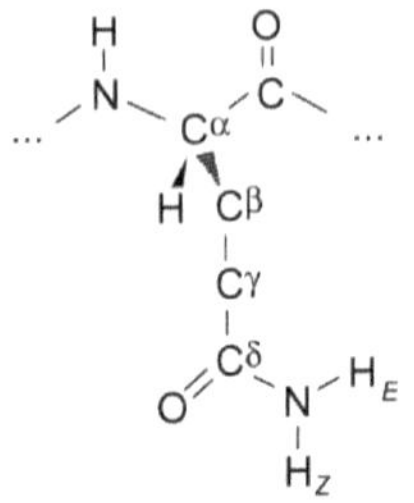

Introduction

$$\ldots \overset{H}{\underset{}{N}} - \underset{\underset{\underset{\underset{O}{\overset{\|}{C^\delta}}-N-H_E}{|}}{\underset{|}{C^\gamma}}}{\underset{\underset{H}{|}}{C^\alpha}} \overset{\overset{O}{\|}}{\underset{}{C}} \ldots$$

Figure 1.1. Example for an amino acid (glutamine) and its nomenclature as used in the following. The side chain carbons are denoted with Greek letters, side chain amide protons are assigned as Z and E for the syn and anti position in respect to the carbonyl oxygen.

The specific shape of the molecule is provided by interaction of amino acids further than defined by the primary sequence of successive peptide bonding. The most important motives are helical winding (α-helix) and flat sheets of β-strands. These are generated by formation of hydrogen bonds (H-bonds) between an amide proton and the oxygen atom of the receptor carbonyl group of the protein backbone. The side chains of residues forming these structural motives point outward (perpendicular in case of the sheets) and may serve for functional aspects of the domain like for the ability of temporary, non-covalent binding to ligands or other proteins e. g.. Figure 1.2 displays a β-strand as an example of a secondary structure element in the SH3 domain of α-spectrin, which is used in this study for methodological purposes and mainly adopts β-sheet conformation.

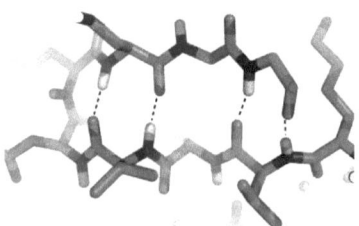

Figure 1.2. Example for a β-strand secondary structural unit. The antiparallel β-sheet is characterized by a close inter-residual contact between amide proton and carbonyl oxygen of different strands (marked by dashed lines). In contrast to a parallel β-sheet, the strands point into opposite directions in an antiparallel one. The depicted β-strand can be found in the SH3-domain of α-spectrin (residues 44-51).

Introduction

These non-covalent bonds form the elements of secondary structure, which can be linked again by covalent (disulfide bonds) or non-covalent interactions (salt bridges, hydrophobic contacts etc.) to generate the tertiary structure of the protein. Being composed of one or multiple domains, each protein adopts a structure which is in close relation to its biological function.

In order for a correct fold of the protein after its primary structure has been generated by the ribosome, diverse mechanisms help to avoid misfolding and a subsequent aggregation. Smaller domains can fold anonymously by an automatic formation of structure motives and tertiary contacts exclusively defined by the primary amino acid sequence. Although an arbitrary permutation of all possible conformations would take a comparably long time,[2] folding is fast due to local initiation of energetically favoured regions (molten globule) driving a successive folding around these parts.[3] Larger proteins are necessary to be prevented from misfolding by a complex machinery of molecular chaperones like the heat shock proteins. A famous chaperonine system is the interplay of GroEL (a member of the Hsp60 family) and GroES, which provide a refolding of unfolded proteins.[4] Another protein of this kind is the "trigger factor", which catches the nascent chain at the exit of the ribosome, where translation into the amino acid sequence is performed.[5,6] This prevents the protein from uncontrolled folding and potential dysfunction.

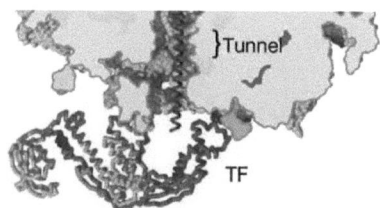

Figure 1.3. Trigger factor (TF) bound to the ribosome. The complex depicted in the picture is formed by the 50S ribosomal subunit of *Haloarcula marismortui*, the TF is from *E. coli*. The picture is taken from Ferbitz et al.[5]

If protein folding is not achieved properly or misfolding happens due to a heat shock e. g., proteins can partly adopt a random coil structure. For a well-folded protein, polar residues tend to stick outwards (vice versa for membrane embedded proteins), while hydrophobic side chains usually point inwards and provide cohesion of the protein core by van-der-Waals forces. For a random coil protein, a random orientation of the backbone amide groups and side chains leads to a potential danger of aggregation. These aggregates can have a pathologic impact on the cell, as supposed for

Introduction

neuro-degenerative illnesses like Huntington[7], Alzheimer's disease[8], or prion diseases.[9] The misfolding of the protein segment $A\beta^{1-40}$ in case of Alzheimer's disease (see a more detailed introduction in Chapter 2.2) is an issue of medical significance and forms the basis for the investigation described in Chapter 5.

Besides a stable scaffold in terms of secondary, tertiary and quaternary structure (which is the interplay of multiple subunits in a protein complex), the flexibility of certain protein parts is an important prerequisite for a great extent of protein function. For this reason, a high interest has prevailed on the elucidation of protein dynamics. Ligand recognition in terms of enzymatic activity e. g. has been thought of as an induced fit process. Conformational changes cannot only be induced by a binding event but also occur in terms of random processes obeying thermodynamic laws. For a recognition event, it has been shown that a kind of conformational preselection plays an additional role.[10] The change of conformation upon influence of physical parameters like temperature or ion concentrations in the solvent can be important for a specificity of processes like the voltage gating of the voltage dependent ion channel VDAC e. g..[11,12] Just like in many other cases, slow dynamics are here the cause for difficulties in the detection of the flexible parts.[12] The processes requiring a dynamic protein fold are enabled by a certain flexibility of covalent bonds, reversible isomerization of proline residues e. g. or changes in the H-bond register. As in case of cooperative binding of oxygen to haemoglobin, conformational changes of the apoprotein can also be induced by cofactors that undergo conformational changes.

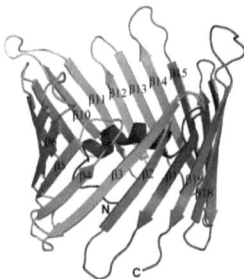

Figure 1.4. Structure of the voltage dependent anion channel VDAC. The rigid β-barrel is bestowed with a mobile central α-helix (dark blue). This flexibility of the helix is supposedly responsible for the specificity of the gating process. The Figure is taken from Bayrhuber et al..[11]

Cellular processes can be observed with manifold different techniques. These focus e. g. on binding events, like surface plasmon resonance (SPR) or isothermal calorimetry (ITC). Others focus on

Introduction

functional analyses, like different essays or patch clamp techniques for processes occurring beyond membranes. The structure and interaction of diverse (bio-) molecules on the atomic level can be probed by manifold spectroscopic methods. For some of these techniques, changes on the atomic level can also be followed in vivo. In living cells, localization of proteins and cellular compartments can also be elucidated by microscopic methods in combination with specific labelling of the respective biomolecules. Interference with the procession of genetic information is even used for observation of the according impact onto a whole organism.

Structural investigation of cellular compartments has reached a level, where microscopic and spectroscopic methods can be used in a combined fashion, providing high resolution information of complex molecular ensembles. X-ray crystallography is the most prominent technique for structure elucidation, having provided most of the 50.000 structures listed in the protein data bank with a high accuracy. Structural investigation of multiple participants of a specific cellular process can lead to a detailed understanding of complex processes. This is the case for e. g. transcription and translation of genetic information,[13] the mechanisms of photosynthesis in plants[14] or for G-protein coupled receptors important for signalling processes like human vision.[15,16]

Figure 1.5. Structure of a G-protein coupled receptor obtained by X-ray crystallography. The depicted structure of the β_2-adrenergic receptor (blue) is shown as a fusion protein with T4 lysozyme (grey) replacing an intracellular helix for a facilitated crystallization. The receptor binds adrenaline and noradrenalin in the course of cardio-vascular and pulmonary regulation and is, like most other GPCRs, a very important drug target. The Figure was taken from Rosenbaum et al.[16]

Nuclear Magnetic Resonance (NMR) spectroscopy has been used as a complementary tool for the elucidation of protein structures.[17,18] Assignment of the individual resonances of spins in a magnetic field can be provided by multidimensional experiments correlating different spatially close nuclei.[19] Distance restraints can then be obtained by correlation experiments in which the peak intensity scales with the proximity of the involved spins.[20] In addition, restraints can be obtained

Introduction

from dihedral angles[21,22] or H-bond detection.[1,23] A high accuracy can be obtained on a basis of a network of individual restraints. This local structural information can be complemented by information about a global orientation of different protein parts based on a partial alignment of the biomolecule in the magnetic field.[24] NMR spectroscopy becomes challenging for molecules of high molecular weight. In comparison to other techniques for structural elucidation like crystallography, however, information about mobility[18,25] or time-dependent processes[26] are obtainable. While sensitivity traditionally requires an extensive concentration of the target molecule, NMR has also provided insights into living cells.[27,28] Due to an inherent insolubility and/or poor crystallization characteristics of many proteins like membrane proteins or disordered proteins, these have been structurally investigated using Magic Angle Spinning (MAS) solid state NMR. This technique is gaining increasing popularity and provides a growing number of entries in the protein data bank.[29-31] Figure 1.6. depicts a structural model of the prion form of the fungal protein HET-s, obtained by solid state NMR of the protein fibrils.[30]

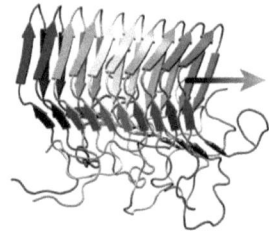

Figure 1.6. Structural model of the HET-s (218–289) fibril published by Wasmer et al.[30] The structure was obtained by MAS solid state NMR of uniformly $^{13}C,^{15}N$ labelled fibrils of the prion protein of filamentous fungus *Podospora anserine*. In addition to well structured β-sheet regions, the N- and C-terminus as well as an intermediate loop region are not resolved.

1.2 The physics of NMR and its quantum-mechanical description[32-34]

Dependent on their spin quantum number *I*, most nuclei comprise a magnetic moment $\vec{\mu}$ going back to a spin angular momentum $\vec{I}$ and a nucleus specific proportionality constant, the gyromagnetic ratio γ.

$$\vec{\mu} = \gamma \vec{I} \tag{1}$$

Introduction

The vectorial momentum has a magnitude

$$|\vec{I}| = \hbar\sqrt{I(I+1)}. \tag{2}$$

Additionally, spin angular momentum $\vec{I}$ can adopt different orientations relative to an external magnetic field B_0, which by definition is oriented in z-direction. This is characterized by the component I_z, which is proportional to the spin quantum number m adopting values of I, $I-1$, ..., $-I$. For a spin ½ particle, $m = ½$ or $-½$. These two states are usually called the α- and β-state, respectively. This also gives us the z-component of the magnetic moment μ_z.

$$\mu_z = \gamma I_z = \gamma \hbar m, \tag{3}$$

which, in interaction with the external static magnetic field B_0, results in an energy E:

$$E = -\mu_z B_0 = -\gamma \hbar m \cdot B_0 \tag{4}$$

For spin ½ nuclei, the resulting energy difference ΔE between the two states is of magnitude $\Delta E = \gamma \hbar B_0$.

Thus, when brought into a magnetic field, spins with $I > 0$ adopt a certain preferential orientation according to Boltzmann. According to Planck, transitions between the states can be affected by (radio-frequency) pulses obeying

$$\omega_{rf} = \omega = \Delta E / \hbar = \gamma B_0, \tag{5}$$

with ω and ω_{rf} being the nuclear resonance (Larmor) frequency and the frequency of the respective pulse, which is also expressed in terms of ν and ν_{rf}, by division by 2π.

Due to relatively small energetic differences between the α- and β-state, however, the according population differences responsible for an overall magnetization is on the order of $1/10^5$ for B_0 field of 11.7 T at room temperature. This renders NMR an insensitive method compared with spectroscopy focusing on electronic states and has led to the construction of strong magnets.

As a characteristic property being different for different spins, the so-called chemical shift (in units ppm) is defined relative to a reference as

$$\delta = (\sigma - \sigma_{ref}) \cdot 10^6. \tag{6}$$

σ is the average isotropic shielding constant, which renders the resonance frequency of a nucleus dependent on its characteristic molecular surrounding:

$$\omega = -\gamma(1-\sigma)B_0 \tag{7}$$

Introduction

Being a non-isotropic constant, its different spatial components σ_{11}, σ_{22}, and σ_{33} are used in order to describe the anisotropy of the chemical shift (CSA), defined by

$$\Delta\sigma = \sigma_{11} - (\sigma_{22} + \sigma_{33})/2 \tag{8}$$

and an asymmetry defined by

$$\eta = \frac{3(\sigma_{22} - \sigma_{33})}{2\Delta\sigma}. \tag{9}$$

For a fast molecular reorientation like in solution, the orientation dependent components average out, such that exclusively the isotropic chemical shift remains.

In order for a quantitative evaluation of spin dynamics, the time dependent wave function $\psi(t)$ of the system can be analyzed by application of operators yielding spin states, energies etc. in terms of a quantum mechanical description.

The time dependence of the wave function is described by the time-dependent Schrödinger equation

$$\frac{d}{dt}|\Psi(t)\rangle = -i\hat{H}(t)|\Psi(t)\rangle, \tag{10}$$

where $\hat{H}$ is the Hamilton operator, defined by $\hat{H}\Psi(t) = E\Psi(t)$ with the eigenvalue E being the energy of the wave function and assuming units in which $\hbar = 1$. The scalar product in Hilbert space between two wave functions χ and ψ is defined as

$$\langle\chi|\psi\rangle = \int \chi^* \psi \, d\tau \quad, \tag{11}$$

using the integrand τ for integration over the whole space. The wave function can be expanded in terms of an orthonormal basis set of basis functions $\{|\psi_i\rangle, i = 1, 2...n\}$ with $\langle\psi_i|\psi_j\rangle = \delta_{i,j}$, meaning

$$|\Psi(t)\rangle = \sum_{i=1}^{n} c_i(t)|\psi_i\rangle \quad. \tag{12}$$

$\delta_{n,m}$ is the Kronecker delta.

The effect that an arbitrary operator $\hat{A}$ has on the wave function is hence determined by variable coefficients $c_i(t)$.

Introduction

$$\left\langle \hat{A} \right\rangle = \int \Psi^*(t)\, \hat{A}\, \Psi(t) = \sum_{i,j} c_i^* c_j \int \psi_i^* \hat{A} \psi_j\, d\tau = \sum_j c_j^* c_j \lambda_j \tag{13}$$

λ_j is the eigenvalue of the operator $\hat{A}$ for the eigenfunction ψ_j.

The different populations and coherences in an ensemble of spins are commonly described via the spin density matrix σ. This matrix is derived from the effect of an arbitrary operator $\hat{A}$ onto the overall average over an ensemble (denoted by an overbar). Starting from (13), but using time independent basis functions $|m\rangle$ and $|n\rangle$ and their variable coefficients c_n and c_m, we obtain:

$$\left\langle \hat{A} \right\rangle = \sum_{nm} c_n c_m^* \left\langle m | \hat{A} | n \right\rangle \tag{14}$$

We can consider $c_n c_m^*$ an element of an operator $\hat{P}$, hence

$$\left\langle \hat{A} \right\rangle = \sum_{nm} \left\langle n | \hat{P} | m \right\rangle \left\langle m | \hat{A} | n \right\rangle = \sum_m \left\langle n | \hat{P}\hat{A} | n \right\rangle = Tr\{\hat{P}\hat{A}\} \tag{15}$$

denoting the trace of the product operator $\hat{P}\hat{A}$. For the ensemble average, we define

$$\overline{c_n c_m^*} = \overline{\left\langle n | P | m \right\rangle} = \left\langle n | \sigma | m \right\rangle = \sigma_{nm} \tag{16}$$

This leads to the similarity

$$\left\langle \overline{A} \right\rangle = Tr\{\sigma \hat{A}\} = Tr\{\hat{A}\sigma\} \tag{17}$$

In order to determine the evolution of spin precession in the magnetic field, the time-dependent Schrödinger-equation can be used to develop the so called Liouville-von-Neumann-equation. For a constant Hamiltonian, this relation describes the evolution of the density operator σ to be the commutator of $\sigma(t)$ and the Hamiltonian $\hat{H}$:

$$\frac{d\sigma(t)}{dt} = -i\left[\hat{H}, \sigma(t)\right] \tag{18}$$

The solution to this differential equation is

$$\sigma(t) = \exp(-i\hat{H}t)\, \sigma(0)\, \exp(i\hat{H}t), \tag{19}$$

which accords to a simple rotation matrix.

The spin states $|\alpha\rangle$ and $|\beta\rangle$ can be expressed in terms of an orthonormal basis set:

Introduction

$$\alpha = \begin{bmatrix} 1 \\ 0 \end{bmatrix}, \quad \beta = \begin{bmatrix} 0 \\ 1 \end{bmatrix} \tag{20}$$

The matrix representations of the spin operators in this basis set can be defined as

$$\hat{I}_x = \frac{1}{2}\begin{bmatrix} 0 & 1 \\ 1 & 0 \end{bmatrix}, \quad \hat{I}_y = \frac{1}{2}\begin{bmatrix} 0 & -i \\ i & 0 \end{bmatrix}, \quad \hat{I}_z = \frac{1}{2}\begin{bmatrix} 1 & 0 \\ 0 & -1 \end{bmatrix} \tag{21}$$

They satisfy the commutation relation

$$\left[\hat{I}_x, \hat{I}_y\right] = i\hat{I}_z . \tag{22}$$

A different basis, which is convenient for the description of coherences rather than the effect of pulses e. g., is the shift operator basis. The operators forming this basis can be derived from the Cartesian basis by the following operations:

$$\hat{I}^+ = \hat{I}_x + i\hat{I}_y = \begin{bmatrix} 0 & 1 \\ 0 & 0 \end{bmatrix}, \quad \hat{I}^- = \hat{I}_x - i\hat{I}_y = \begin{bmatrix} 0 & 0 \\ 1 & 0 \end{bmatrix}$$

$$\hat{I}_0 = \sqrt{2}\,\hat{I}_z = \frac{1}{\sqrt{2}}\begin{bmatrix} 1 & 0 \\ 0 & -1 \end{bmatrix}, \quad \frac{1}{\sqrt{2}}\hat{E} = \frac{1}{\sqrt{2}}\begin{bmatrix} 1 & 0 \\ 0 & 1 \end{bmatrix} \tag{23}$$

Knowledge about the density matrix in the course of an experiment will tell us about the observable magnetization that we have to expect during acquisition. For a tailored evolution of NMR coherences, the Hamiltonian can be modified to consist of free evolution delays on the one hand and radio-frequency irradiation in order for short pulses, recoupling of interactions etc. on the other hand.

During free evolution, the Zeeman interaction leads to precession of spin angular momentum around the axis of the interaction, which is the magnetic field B_0:

$$\hat{H} = \omega \hat{I}_z$$

According to (19), the evolution of the density operator upon precession of the spins in the B_0-field can be described by a simple rotation around the axis of the interaction. Starting with a density matrix $\sigma = -\hat{I}_y$, e. g. (which would be the case after a 90° x-pulse), the precession will lead to

Introduction

$$\sigma(t) = \exp(-i\hat{H}t)\,\sigma(0)\exp(i\hat{H}t)$$
$$= -\exp(-i\omega \hat{I}_z)\hat{I}_y \exp(i\omega \hat{I}_z) \quad (24)$$
$$= \hat{I}_x \sin(\omega t) - \hat{I}_y \cos(\omega t)$$

In order to simplify the Hamiltonian, it is mostly transformed to the rotating frame. This reduces the description of the precession of nuclear spins in the static magnetic field B_0 to the small deviation of the Larmor frequency due to the characteristic electronic environment. As a similar case, the application of radio-frequency pulses can be described by the rotation around an effective rotation axis. The effective (time independent) Hamiltonian for an on-resonance pulse with phase x,

$$\hat{H}_e = \omega \hat{I}_x \quad (25)$$

can be obtained by application of a unitary transformation $U = \exp(i\omega_{rf}\hat{I}_z t)$ to the time dependent Hamiltonian

$$\hat{H} = \hat{H}_z + \hat{H}_{rf} = \omega_0 \hat{I}_z + \omega_1[\hat{I}_x \cos(\omega_{rf}t + \phi) + \hat{I}_y \sin(\omega_{rf}t + \phi)]. \quad (26)$$

The respective rotation of σ around the axis of interaction can be shown to be (at time point τ_p):

$$\sigma(\tau_p) = -\exp(-i\alpha \hat{I}_x)\sigma(0)\exp(i\alpha \hat{I}_x), \quad (27)$$

with

$$\exp(-i\alpha \hat{I}_x) = \begin{bmatrix} \cos(\alpha/2) & i\sin(\alpha/2) \\ i\sin(\alpha/2) & \cos(\alpha/2) \end{bmatrix} \text{ and } \exp(i\alpha \hat{I}_x) = \begin{bmatrix} \cos(\alpha/2) & -i\sin(\alpha/2) \\ -i\sin(\alpha/2) & \cos(\alpha/2) \end{bmatrix} \quad (28)$$

The effect onto an $\hat{I}_z$-operator, e. g., would be the rotation around the x-axis by an angle α:

$$\sigma(\tau_p) = -\exp(-i\alpha \hat{I}_x)\hat{I}_z \exp(i\alpha \hat{I}_x)$$
$$= \hat{I}_z \cos\alpha - \hat{I}_y \sin\alpha \quad (29)$$

Additionally, couplings to other nuclei or (unpaired electrons) can occur, such as scalar couplings (J-couplings to covalently bonded nuclei through the electrons in the bond) and dipolar couplings, which are anisotropically transferred through space. In order to treat these phenomena, a two (multiple) spin system has to be considered. The extension of a one spin wave function is usually done via the direct product operation:

$$\Psi_k = |m_1\rangle \otimes |m_2\rangle \otimes ... \otimes |m_n\rangle = |m_1, m_2, ..., m_n\rangle \quad (30)$$

The direct product of matrices can be illustrated for a (2 x 2) matrix, being:

Introduction

$$A \otimes B = \begin{pmatrix} A_{11}B & A_{12}B \\ A_{21}B & A_{22}B \end{pmatrix} = \begin{pmatrix} A_{11}B_{11} & A_{11}B_{12} & A_{12}B_{11} & A_{12}B_{12} \\ A_{11}B_{21} & A_{11}B_{22} & A_{12}B_{21} & A_{12}B_{22} \\ A_{21}B_{11} & A_{21}B_{12} & A_{22}B_{11} & A_{22}B_{12} \\ A_{21}B_{21} & A_{21}B_{22} & A_{22}B_{21} & A_{22}B_{22} \end{pmatrix} \quad (31)$$

The *J*-coupling Hamiltonian for N (= 2) scalar coupled spins is

$$\hat{H} = 2\pi J_{i,j} \sum_{i=1}^{N-1} \sum_{j=i+1}^{N} \hat{I}_i \cdot \hat{S}_j \quad \stackrel{N=2}{(=} 2\pi J_{I,S} \hat{I} \cdot \hat{S}) \quad (32)$$

For most cases, the weak coupling condition $2\pi J_{I,S} |\omega_i - \omega_j| \ll 1$ is fulfilled and the Hamiltonian can be simplified to be

$$\hat{H} = 2\pi J_{I,S} \hat{I}_z \hat{S}_z \quad (33)$$

Evolution of the density matrix under the scalar coupling Hamiltonian results in

$$\sigma(t) = \exp(-i\alpha 2\ I_z S_z)\sigma(0)\exp(i\alpha 2\ I_z S_z) \qquad \text{with } \alpha = \pi J_{I,S}\ t \quad (34)$$

Since

$$\exp(i\alpha 2\ \hat{I}_z \hat{S}_z) = E \cos\frac{\alpha}{2} + 4i\hat{I}_z \hat{S}_z \sin\frac{\alpha}{2}, \quad (35)$$

with *E* being the unity matrix of a dimension (4x4), the propagator for free evolution (including chemical shift evolution) of the two-spin system is

$$e^{i(\Omega_I I_z + \Omega_S S_z + 2\pi J_{I,S} I_z S_z)t}$$

$$= \frac{i}{2} \begin{bmatrix} e^{i(\Omega_I + \Omega_S + \pi J_{I,S})t} & 0 & 0 & 0 \\ 0 & e^{i(\Omega_I - \Omega_S - \pi J_{I,S})t} & 0 & 0 \\ 0 & 0 & e^{i(-\Omega_I + \Omega_S - \pi J_{I,S})t} & 0 \\ 0 & 0 & 0 & e^{i(-\Omega_I - \Omega_S + \pi J_{I,S})t} \end{bmatrix} \quad (36)$$

After performance of the matrix multiplication with the density operator and a multiplication with

$$F^+ = \begin{bmatrix} 0 & -1 & -1 & 0 \\ 0 & 0 & 0 & -1 \\ 0 & 0 & 0 & -1 \\ 0 & 0 & 0 & 0 \end{bmatrix} \propto I^+ + S^+, \quad (37)$$

the *observable* magnetization is described by:

$$\langle M^+ \rangle(t) = const. \cdot (e^{i(\Omega_I + \pi J_{I,S})t} + e^{i(\Omega_I - \pi J_{I,S})t} + e^{i(\Omega_S + \pi J_{I,S})t} + e^{i(\Omega_S - \pi J_{I,S})t}) \quad (38)$$

Introduction

The effect of the coupling is a splitting of each resonance frequency with separation $J_{1,2}$.

For a more practical description of the spin system, one often sticks to the operator of interest itself instead of the density matrix. This approach, which can be used in case of weakly coupled spins, is called the product operator formalism.[35] In this formalism, simple basic rules allow to predict the outcome of an experiment for most (scalar transfer based) pulse sequences.

For the Zeeman interaction, e. g.,

$$\hat{I}_x \xrightarrow{\Omega_I I_z t} \hat{I}_x \cos(\Omega_I t) + \hat{I}_y \sin(\Omega_I t) \tag{39}$$

similarly, for pulses

$$\hat{I}_z \xrightarrow{\alpha I_{\pm x}} \hat{I}_z \cos\alpha \pm \hat{I}_y \sin\alpha \tag{40}$$

and for J-coupling evolution:

$$\hat{I}_x \xrightarrow{2\pi J_{I,S} I_z S_z t} \hat{I}_x \cos(2\pi J_{I,S} I_z t) + 2\hat{I}_y \hat{I}_z \sin(2\pi J_{I,S} I_z t). \tag{41}$$

These rules are based on cyclic permutations of spin operators (of the same kind):

$$[\hat{A}, \hat{B}] = i\hat{C}. \tag{42}$$

In contrast to in-phase operators like $\hat{I}_x$ and $\hat{I}_y$, antiphase operators like $\hat{I}_y \hat{I}_z$ can not be detected directly. However, scalar magnetization transfer bases on the generation of such operators. In order to fully evolve anti-phase operators like $\hat{I}_y \hat{I}_z$, the approximate magnitudes of J-couplings is crucial to know. Respective values are depicted for the protein backbone in Figure 1.7.

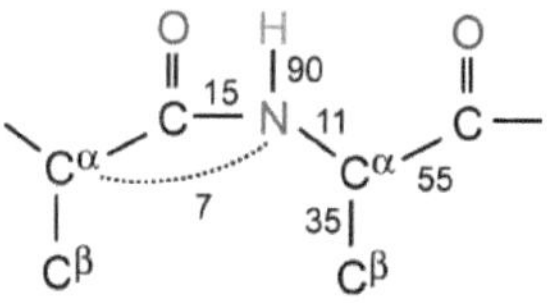

Figure 1.7. Approximate heteronuclear 1- and 2-bond scalar couplings (in Hz) of the protein backbone. The values can vary slightly for different backbone geometry or secondary structure conformation.[36]

As one of the most important building blocks, the magnetization transfer via scalar couplings in terms of INEPT (Insensitive Nuclei Enhanced Polarization Transfer)[37] enables signal enhancement due to use of convenient starting and detection nucleus (in terms of relaxation properties and their

Introduction

gyromagnetic ratio γ). In general, it provides a tool extensively used for heteronuclear multidimensional NMR spectroscopy. In order for a tailored NMR experiment, one makes use of an innumerable pool of pulse sequence elements effecting the spin dynamics in the way described above.

1.3 Relaxation[33]

The coherences formed by tailored application of radio-frequency pulses do not represent the equilibrium Boltzmann distribution of spin states in the sample. By fluctuations of the effective field felt by each nuclei, the equilibrium state is restored by successive spin flips. These fluctuations may result from an overall reorientation of the molecule, internal dynamics, motion of a paramagnetic center relative to the nucleus etc. The quantum-mechanical description of the observed relaxation is beyond the scope of the introductory part of the thesis. However, the mechanisms participating in the observable relaxation properties of biomolecules are discussed briefly.

Going back to the Bloch, Wangsness and Redfield theory, the spectral density function

$$j^q(\omega) = \text{Re}\left\{\int_{-\infty}^{\infty}\overline{F_k^q(t)F_k^{-q}(1+\tau)}\exp(-i\omega\tau)d\tau\right\} \tag{43}$$

is derived from a stochastic Hamiltonian $\hat{H}_1(t) = \sum_{q=-k}^{k}(-1)^q F_k^{-q}(t)\hat{A}_k^q$ with a random function of spatial variables $F_k^q(t)$ and tensor spin operator A_k^q, with k being the rank of the tensor. For rotational diffusion of a spherical particle, the density function can be dissected into an orientational correlation function $J(\omega)$ and a term d_{00} including the physical constants relevant for the considered interaction. Solving the integration of the correlation function for a random stochastic process like overall reorientation in solution e.g.,

$$j(\omega) = d_{00} \cdot J(\omega) \quad \text{with } J(\omega) = \frac{2}{5}\frac{\tau_c}{1+\omega^2\tau_c^2}. \tag{44}$$

Dipolar relaxation

In large biomolecules, relaxation of spin ½ is usually determined by chemical shift anisotropy and dipole-dipole interactions. (Nuclei with a spin larger than ½, which in biomolecules are almost exclusively deuterons and ^{14}N, are relaxed in particular by quadrupolar interactions.) Dipolar re-

Introduction

laxation goes back to dipole-dipole interactions as described below. For populations and coherences, a contribution to the relaxation of nucleus I by spin S can be described by a longitudinal rate

$$R_1^{DD} = (d_{00}/4)\{J(\omega_I - \omega_S) + 3J(\omega_I) + 6J(\omega_I + \omega_S)\} \tag{45}$$

and a transverse rate

$$R_2^{DD} = (d_{00}/8)\{4J(0) + J(\omega_I - \omega_S) + 3J(\omega_I) + 6J(\omega_S) + (6J(\omega_I + \omega_S)\} \tag{46}$$

(given for single quantum coherence I^+), respectively. For a complete description see Cavanagh[33] etc. The dipolar coupling constant

$$d_{00} = \frac{\mu_0^2 \hbar^2 \gamma_I \gamma_S}{(4\pi)^2 r_{IS}^6} \tag{47}$$

includes the distance dependency of $1/r^6$. Through a further dependency on γ the amount of dipolar relaxation strongly reflects the abundance of protons.

CSA induced relaxation

Chemical Shift Anisotropy (CSA) describes the orientation dependency of chemical shifts in respect to the external magnetic field. The longitudinal and transverse relaxation rates going back to CSA can be deduced as

$$R_{1I}^{CSA} = d_{00} J(\omega_I) \quad \text{and} \tag{48}$$

$$R_{2I}^{CSA} = (d_{00}/6)\{4J(0) + 3J(\omega_I)\} \tag{49}$$

with $\quad d_{00} = (\Delta\sigma \gamma_I B_0)^2/3 = \Delta\sigma \omega_I^2/3$. \hfill (50)

The generalized CSA constant $\Delta\sigma$ is dependent on the anisotropy and asymmetry of the respective chemical bonds. The dependency on B_0 induces a strong CSA relaxation for NMR at large magnetic fields. CSA induced relaxation is most important for nuclei with large chemical shift ranges like ^{15}N, ^{13}C, ^{31}P etc.

Relaxation interference

Introduction

As deduced elsewhere,[33] dipolar and CSA relaxation add constructively and destructively for the different multiplet components of the signal of a nucleus J-coupled to another spin:

$$\frac{d}{dt}\begin{bmatrix}\langle I^+ S^\alpha\rangle(t)\\ \langle I^+ S^\beta\rangle(t)\end{bmatrix} = -\begin{bmatrix} i\pi J_{IS}+\overline{R}_2+\eta_{xy} & (R_{2I}^{DD}-R_{2IS}^{DD})/2 \\ (R_{2I}^{DD}-R_{2IS}^{DD})/2 & -i\pi J_{IS}+\overline{R}_2-\eta_{xy}\end{bmatrix}\begin{bmatrix}\langle I^+ S^\alpha\rangle(t)\\ \langle I^+ S^\beta\rangle(t)\end{bmatrix} \quad (51)$$

with $\quad \eta_{xy} = \frac{\sqrt{3}}{6} cd P_2(\cos\theta)\{4J(0)+3J(\omega_I)\}$ (interference term), (52)

$$c = \gamma_I(\sigma_\parallel - \sigma_\perp)B_0/\sqrt{3}, \quad (53)$$

$$d = -(\mu_0 \hbar \gamma_I \gamma_S)/(4\pi r_{IS}^3), \quad \text{and} \quad (54)$$

$$\overline{R}_2 = (R_{2I}^{DD}+R_{2IS}^{DD})/2 + R_{2I}^{CSA}. \quad (55)$$

R_{2I}^{DD} and R_{2IS}^{DD} refer to dipole-dipole relaxation of in-phase and anti-phase coherence, respectively. P_2 refers to the Legendre-polynom of order 2. The off-diagonal elements can usually be neglected if $\left|(R_{2I}^{DD}-R_{2IS}^{DD})\right| \ll \left|i\pi J_{IS}\right|$.

The effect is coined relaxation interference and has been exploited e. g. for mutual cancellation of dipolar and CSA relaxation in HN spin pairs, resulting in comparably long-lived coherences and sharp linewidths especially for large biomolecules at high magnetic fields. This effect comes into play in the experiments exploiting the TROSY pulse scheme, which are shown in Chapter 4.3.

1.4 Traditional solid state NMR[34]

In solution, rapid tumbling (reorientation of the molecule) leads to an effective averaging of anisotropic interactions, as explained for the chemical shift (vide ultra). In the absence of this rotational diffusion, direct spin-spin interaction by their nuclear dipole moment (dipolar coupling) through space is non-negligible. For two spins, the dipolar coupling Hamiltonian is:

$$H^{DD} = \frac{\mu_0}{4\pi}\hbar\frac{\gamma_1\gamma_2}{r_{1,2}^3}\left(\hat{I}_1\cdot\hat{I}_2 - \frac{3(\hat{I}_1\cdot\vec{r}_{1,2})(\hat{I}_2\cdot\vec{r}_{1,2})}{r_{1,2}^2}\right), \quad (56)$$

in which $b_{1,2} = \frac{\mu_0}{4\pi}\hbar\frac{\gamma_1\gamma_2}{r_{1,2}^3}$ is usually abbreviated as the dipolar coupling constant. For the internuclear vectors expressed in polar coordinates, the spherical representation in terms of irreducible tensor operators yields:

Introduction

$$H^{DD} = \sum_{q=-2}^{2} F_{1,2}^{(q)} A_{1,2}^{(q)} \tag{57}$$

with $F_{1,2}^{(q)}$ and $A_{1,2}^{(q)}$ describing the orientation and the spin operators.

In the high-field approximation, it is possible to neglect the non-secular contributions with $q \neq 0$, which yields:

$$H^{DD} = (1 - 3\cos^2 \theta_{1,2}) \cdot \frac{1}{2} b_{1,2} \left(3\hat{I}_{1z}\hat{I}_{2z} - \hat{I}_1\hat{I}_2\right) \tag{58}$$

For the case of heteronuclear spins I and S, all terms involving transverse operators can be dropped, yielding

$$H^{DD} = b_{1,2}(1 - 3\cos^2 \theta_{1,2})\hat{I}_z\hat{S}_z \tag{59}$$

In both cases, however, the angle dependent term $3\cos^2 \theta - 1$ averages out in case of a stochastically varied value for θ. Thus, in solution state NMR dipolar interactions between the nuclei can generally be disregarded. They have to be taken into account only if the sample does not tumble quickly in respect to the magnetic field. This is the case e. g. for an alignment in the course of RDC (Residual Dipolar Couplings) measurements.[38,39] Apart from that, dipolar interactions induce relaxation of the NMR coherences and populations (vide ultra), leading to faster relaxation and extensive deuteration strategies for large molecules with long reorientation times τ_c.

In the solid state, a motional averaging can be induced artificially employing a rotation around only one axis. Therefore, the respective angle θ between the axis of rotation and B_0 is set to the value where the spatial term is averaged out, meaning $1 - 3\cos^2 \theta = 0$. This angle ($\theta = 54.74°$) is called the *magic angle*. The respective spectroscopic approach is coined magic angle spinning (MAS) and has become a standard method in ssNMR[29,40,41]. Also chemical shifts are affected by the orientation of the bond vector between nuclei in respect to B_0. This anisotropy is also effectively averaged out by a fast rotation around the magic angle.

If needed, dipolar couplings as well as the CSA interaction can be reintroduced in solid state NMR, most commonly by manipulation of the spin part of the Hamiltonian. Examples for selective dipolar recoupling methods are Rotational Resonance (R^2),[42] Radio Frequency Driven Recoupling (RFDR),[43,44] Rotational Echo Double Resonance (REDOR),[45] and Cross Polarization (CP),[46] e. g..

Introduction

In accordance to the use of *J*-couplings in solution, these methods are commonly used for a magnetization transfer among nuclei of interest.

Most important for the work described in this manuscript, CP recouples the spins of interest by their dipolar interaction upon radio-frequency irradiation on both nuclei. The difference of the effective fields on the two channels $\Delta\omega_{rf}$ must match the rotation frequency ω_r in such a way that the Hartmann-Hahn condition[47] for rotating solids is fulfilled, meaning that

$$\Delta\omega_{rf} = n \cdot \omega_r, \tag{60}$$

with n = -2, -1, 1, 2. As derived elsewhere (see Duer[34] e. g.), this resonance leads to an effective energy conserving contact in the doubly rotating frame of interaction between the different nuclei. For a directed CP, which transfers magnetization only between selected nuclei, SPECIFIC-CP (SPECtrally Induced Filtering In Combination with Cross Polarization) has been used.[48] This variant allows for specific N-CO or N-C$^\alpha$ transfers, respectively, by matching the fields such that a specific offset of the irradiation results in a mismatch for the unwanted transfer.

Rotation speed usually found in MAS solid state NMR is on the order of 10 to 50 kHz. At these frequencies, dipolar interactions involving heteronuclei as well as chemical shift anisotropy can be averaged out effectively. Dipolar couplings to ^{1}H, however, can not be averaged out effectively even by fast MAS. Therefore, much work is dedicated to provide decoupling sequences with high efficiency without strong heating effects. ^{1}H-^{1}H dipolar interaction are of the size of approximately 130 kHz at a distance of 1 Å. MAS alone is thus not able to circumvent the difficulties of ^{1}H-detection. Decoupling, as in case of acquisition of heteronuclei, seems difficult for concomitant NMR detection. Thus, traditional MAS solid state NMR relies on detection of heteronuclei and strong heteronuclear decoupling.

1.5 Proton chemical shifts in solid state NMR

Protons have played an important role for indirect magnetization transfer through their strong dipolar coupling network. This is reflected in many mixing sequences using the proton-proton interaction as a measure for distance restraints, like Dipolar Assisted Resonance Recoupling (DARR),[49] DREAM[50] and many others (see Chapter 4.5). Even for a directed N/C-transfer, the proton dipolar interactions can be employed, as in case of Proton Assisted Insensitive Nuclei Cross Polarization (PAIN).[51] However, acquisition of proton chemical shifts is difficult by conventional methods.

Introduction

Indirect ^{1}H detection at comparably low spinning frequency has been achieved in the past by use of windowed homonuclear decoupling schemes.[52-54] These techniques employing frequency-switched Lee-Goldberg decoupling (FSLG)[52] or phase modulated Lee-Goldberg decoupling (PMLG)[53] for keeping the magnetization of protons in the rotating frame constantly along the magic angle, e. g.. In comparison to the omittance of decoupling, the relaxation is largely reduced. However, the effective field on protons is scaled down by a factor of ~1.5, which compromises the obtainable resolution and sensitivity. For high spinning frequencies, proton acquisition has even been performed without homonuclear decoupling.[55] This allows for a direct acquisition, which increases the signal to noise in comparison to detection of low-gamma nuclei X by approximately $(\gamma_H/\gamma_X)^{3/2}$. However, achievable proton linewidths are as broad as ~400 Hz in case of 40 kHz MAS,[55] which compromizes both, resolution and sensitivity.

A decrease of the proton content by specific labelling strategies resulted in severely reduced linewidths.[56] Deuteration of peptides with a back-exchange of amide protons at moderate spinning speeds resulted in a proton linewidth of already 150 Hz.[57] This approach was shown to yield a signal enhancement of a factor 5 and ~10 for 13 and >30 kHz MAS in comparison to ^{15}N detection. At higher spinning frequencies, back-exchange of labile sites in deuterated solids has been used for assignment experiments and structure determination of the small model protein ubiquitin.[58] Achievable linewidths at 40 kHz MAS amount to around 140 Hz and 40 Hz for ^{1}H and ^{15}N, respectively, leading to triple resonance experiments with impressive signal to noise.

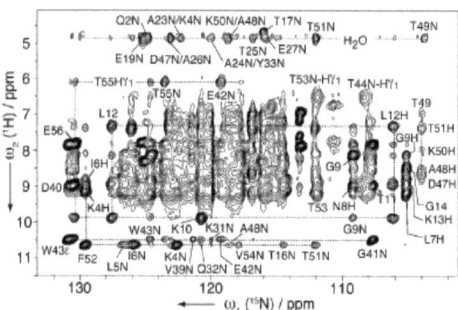

Figure 1.8. ^{1}H detected N/H 2D correlation of ubiquitin with an RFDR mixing step of 3 ms. In addition to direct amide contacts, also through space correlations to other amide or water protons appear in the spectra. The resolution already enables a clear separation of different ^{1}H resonances in many cases. The picture was taken from Zhou et al.[58]

Introduction

Further dilution of the proton content has been used by Chevelkov et al..[59] The reduction of ^{1}H content in exchangeable sites with an otherwise complete deuteration provided linewidths as narrow as ~15-20 Hz at moderate spinning frequencies. Thus, the reduction of NMR active nuclei by a factor of 10 is compensated by the narrowing of the line in addition to a highly increased resolution. The obtainable relaxation properties almost reach solution state conditions. Therefore, scalar transfers can be used instead of CP.[60] In principle, a 100 % transfer efficiency would be achieved by the use of INEPT, yielding comparably high signal intensity in correlation experiments. At the same time, a high spectral purity due to exclusive correlation of covalently bonded nuclei is maintained.

A general obstacle of sparse protonation, however, are significantly prolonged longitudinal relaxation times T_1. Due to this, long recycle delays have to be implemented, which generally causes an additional sensitivity loss. Furthermore, in traditional solid state NMR, much useful information is commonly obtained from the carbon atoms of the protein backbone and side chains. These have so far been excluded almost completely. Approaches based on the protonation of methyl groups in combination with scalar ^{13}C mixing have resulted in valuable side chain carbon correlations.[61,62] The respective information, however, is restricted to valine, leucine, and isoleucine, due to the lack of methyl groups in other amino acids. In case of protonation exclusively for H^N, long range magnetization transfer from ^{1}H to ^{13}C uses up a large fraction of the starting magnetization, resulting in poor signal to noise for correlations involving ^{13}C.

In order for a useful implementation of proton solid state NMR, these limitations have to be overcome. Given a significant enhancement of the signal to noise with a maintained resolution, however, solid state NMR will be able to reach the feasibility of solution state NMR without the intrinsic restriction to soluble biomolecules of limited molecular weight.

So far, the applicability of experiments based on extensive proton dilution has still been restricted to model proteins and simple correlation experiments. The objective of the thesis was the assumption that a further signal to noise improvement would allow for valuable multidimensional experiments on these samples. Since a transfer between nuclei suffers much less from concomitant signal decay if dipolar relaxation is reduced, sparse protonation was thought to allow for triple resonance J-based experiments. These were imagined to yield backbone assignments, information concerning the protein dynamics etc. In the long run, also investigation of larger proteins could be facilitated with the inclusion of ^{1}H chemical shifts.

2 PROTEINS AND PREPARATION

2.1 The SH3 domain of chicken α-spectrin as a system for solid state NMR methods development

The main part of the work in this thesis was performed on the Src homology 3 (SH3) domain of chicken α-spectrin. The family of SH3 domains usually plays a role for protein-protein interaction and protein assembly in signal transduction pathways. In the human genome, approximately 300 SH3 domains are encoded. They consist of approximately 65 amino acids and were first found as a consensus sequence in the viral adaptor protein v-Crk.[63,64] The Src (Sarcoma virus) homology refers to the product of the first proto-oncogene to be identified.[65] Epitopes interacting with SH3 domains usually contain proline-rich motives.[66] These bind to the hydrophobic pocket of the SH3 domain.[67] The interaction can take place far away from the catalytic center, as e. g. in tyrosine kinases,[68,69] and provides an increased substrate specificity. As an example, Figure 2.1 displays the SH3 domain of p40[phox] binding p47[phox] containing a class II poly-proline motive.[70] The interaction of the involved proteins plays a regulatory role for the initiation of NADPH oxidase activation, which produces superoxide anions in response to infection.

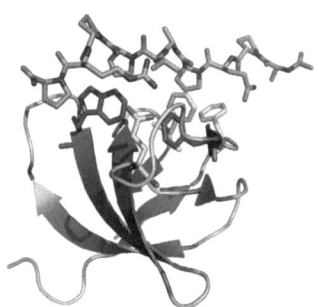

Figure 2.1. Example for an SH3 binding event in the course of NADPH oxidase activation. The polyproline motive of the cytosolic factor p47[phox] (green) binds to the hydrophobic pocket of the SH3 domain of p40[phox] (blue/brown). The picture was generated using the crystal structure 1w70.[70]

Proteins and Preparation

Like ubiquitin, the SH3 domain of chicken α-spectrin[71] has successfully been used as a model system for solid state NMR methods development.[60,72] The SH3 domain is adequately small, easy to produce using *E. coli* strains and enables simple purification with good protein yields. Its ability to form micro-crystals[73] upon pH-shift within a few hours provides NMR samples with a good homogeneity. These features enable its use for methodological investigation and for fundamental biophysical research.

The α-spectrin SH3 fold is typical for SH3 domains and consists of five β-strands in a tight barrel shape. The linker region between P54 and V58 consists of a short α-helix. The secondary structure elements are separated by loops which are coined RT loop, N-Src-loop, and distal loop). Although in the outer parts of the RT-loop no β-sheet structure is recognized by Pymol[74] (see Figure 2.2), a certain rigidity can be assumed here due to interresidual H-bonding.

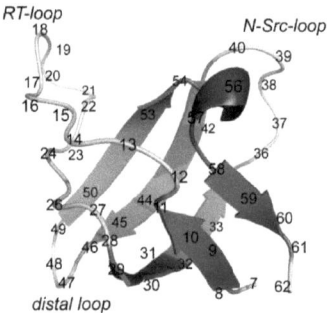

Figure 2.2. Secondary structure of the chicken α-spectrin SH3 domain (pdb-code 2NUZ).[60] The rigid β-sheets (blue) are connected by three loops. A short helix motive can be found between residues 55 and 58 (red).

Recombinant expression of uniformly ^{13}C, ^{15}N, ^{2}D-labelled SH3 was performed in *E. coli* BL21 (DE 3) using a pET3d vector encoding for the protein. Cells were grown in M9 minimal media at 37 °C up to an optical density OD_{600} = 0.8, providing 4g/L u-[^{13}C, ^{2}D] glucose, 1 g/L ^{15}N-NH$_4$Cl and 100% ^{2}D$_2$O.[75,76] Expression was performed over night upon decrease of the temperature to 22 °C and induction by 1 mM IPTG. Cell lyzation via French press, and protein purification via anion exchange chromatography using Q-sepharose FF and gel filtration using Superdex 75 were pursued in Tris buffer containing 100 % H$_2$O at a pH of 3.5. After exchange of the buffer against unbuffered HCl$_{aq.}$ of the same pH and addition of 0.02 % NaN$_3$, the protein solution was lyophi-

Proteins and Preparation

lized. For micro-crystallization, the lyophilisate was resuspended in H_2O/D_2O mixtures of different ratios and followed by additional solvation of 100 mM lyophilized $(NH_4)_2SO_{4aq.}$. After pH-shift of the solution to 7.5 by solvation of NH_3 gas, crystallization was typically accomplished over night.[60] In the course of the thesis, SH3 samples with 10, 20, 25, 30, 40, 60, 80, and 100 % H_2O were produced.

In order to assure an efficient data acquisition, samples were doped by use of paramagnetic ions providing an increased longitudinal relaxation rate R_1 (see below). Numerous examples for the use of relaxation enhancing effects have been reported in the literature and are directed towards local attenuation of signals by elevated transverse relaxation rates as well as an overall reduction of acquisition times.[77-81] For the latter case, T_1 should be decreased as much as possible while transverse relaxation time T_2 would ideally be unaffected. As shown in Chapter 3, employment of paramagnetic doping in ^{1}H-detected MAS solid state NMR allows for cutting down recycle delays principally without restriction.

Micro-crystal quality seems crucial for high resolution NMR spectroscopy in the solid state. Crystal quality upon doping was tested using unlabelled protein produced as described above. In order to assure an ordinary crystallization process, conditions for the pH-shift of the initiation of crystallization were kept close to the established conditions using 5 mg/ml protein and 100 mM $(NH_4)_2SO_4$. A substitution of this salt against paramagnetic ions was tested for chelated Cu^{II}, Ni^{II} and Mn^{II} of a concentration of 50 mM. As a standard chelate, ethylene diamine tetraacetate (edta) was used. For the preparation of the complex, H_4edta was added to the metal sulphate with a 5 % excess and the pH brought to 7 by help of $NH_{3aq.}$. This provided a 1:1 mixture of $(NH_4)_2SO_{4aq.}$ and $(NH_4)_2[M^{II}(edta)]_{aq.}$, which could be obtained in high concentrations (350 mM) using an iterative dissolution of salts and NH_3. Finally, the solution was lyophilized. For high paramagnetic salt concentrations (> 50 mM), additional dissolution of $(NH_4)_2SO_{4aq.}$ was omitted. For low concentration of paramagnetic salts (< 50 mM), $(NH_4)_2SO_{4aq.}$ was substituted such that ionic strengths of the solution were maintained. All crystallization conditions resulted in comparable micro-crystals. Figure 2.3 shows light microscopic pictures of the crystals under indicated conditions.

Proteins and Preparation

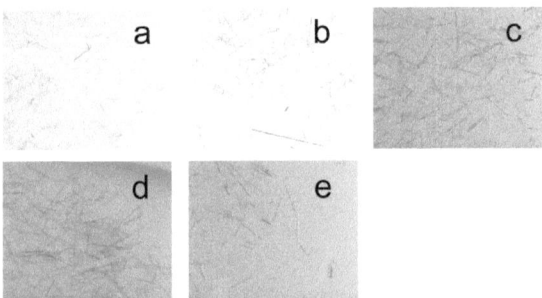

Figure 2.3. Micro-crystals formed under concentrations of chelated cation sulfates, 3 mM Cu^{II} (a), 20 mM Cu^{II} (b), 50 mM Cu^{II} (c), 50 mM Ni^{II} (d), and 50 mM Mn^{II} (e).

For Cu^{II}, different paramagnetic salt concentrations were tested. These did not give rise to significant differences. Crystals formed upon presence of Mn^{II} degraded within a few days in contrast to the other chelates. This process was indicated by an increasingly yellow colour of the solution. For the investigation of NMR spectroscopic properties of Cu^{II} doped samples, concentrations of 20 mM, 75 mM, 150 mM, and 250 mM Cu^{II} were used.

Rotor filling was achieved by centrifugation of the micro-crystal suspension in a pipette tip connected to an NMR rotor and successive centrifugation of the rotor alone. Supernatant buffer was removed with a small piece of tissue.

2.2 Aβ^{1-40}, a fibril forming 40-residue fragment of the Amyloid Precursor Protein APP

The 40-residue peptide Aβ^{1-40} is related to the pathology of the most common age-related neurodegenerative disorder, Alzheimer's disease.[8] This protein fragment arises in the course of the cleavage of the Amyloid Precursor Protein APP, involved in signal transduction through the cell membrane.[82] Upon generation of β-APP by γ-secretase, a 99 amino acid fragment, C99, remains in the membrane. Subsequent APP cleavage by β-secretase BACE (β-site APP cleaving enzyme) releases Aβ as an insoluble product that aggregates and forms fibrillic structures. Besides Aβ^{1-40}, the slightly longer Aβ^{1-42} is generated to a 10 % extent. Aβ^{1-42} is even more prone to form plaques than Aβ^{1-40}.[83] Plaques are not formed if cleavage takes place by α-secretase instead of BACE. In this case, a 83 amino acid long intermediate (C83) is formed instead, which is not prone to aggregation.[84]

Proteins and Preparation

The cause for neurodegerative effects has been shown to go back to different other origins than formation of fibrils and neurofibrillary tangles,[85] like e. g. inflammatory responses[86] and oxidative stress.[87] There has been evidence that the toxic species in the course of the disease are intermediately occurring oligomers rather than fibrils.[88] Nevertheless, a detailed molecular understanding of the amyloid structure is generally desired for a potential improvement of medical treatment against Alzheimer's disease. A structural model of amyloid fibrils based on the published torsion angles[89] is shown in Figure 2.4. There has been evidence for different conformations in respect to the inter-strand assembly of the β-sheets: A refined model by Tycko et al. supposes trigonal geometry in addition to the before-hand published assembly having C_2 symmetry.[90]

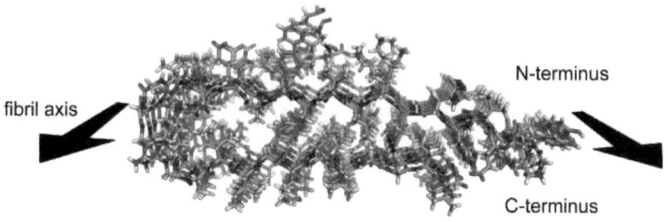

Figure 2.4. Structural model of $A\beta^{1-40}$, according to published torsion angles.[89] Two β-sheets are formed by the N-terminal and the C-terminal residues.

Expression of uniformly [^{2}D, ^{15}N, ^{13}C]-labeled $A\beta^{1-40}$ was achieved by recombinant expression in E. coli (BL21 DE3), using a p28a vector (Novagen) carrying an insert encoding for the $A\beta^{1-40}$ sequence. Expression tests were performed in LB, subsequent expression of labelled protein pursued in isotopically enriched M9 medium (0.5 g/l ^{15}NH$_4$Cl, 2 g/l ^{13}C glucose) containing 50 mg/l kanamycin. Cell cultures were grown to an OD$_{600}$ of 0.6 at 37 °C and induced by 1 mM IPTG. The cells were harvested after 4 h by centrifugation, the pellet was resuspended and lysed by sonication. Inclusion bodies were purified using a differential centrifugation-detergent wash procedure[91,92] with repeated washing (resuspension of the pellet by sonication and centrifugation) in a buffer consisting of 50 mM Tris-HCl pH 7.5, 100 mM NaCl, 1 mM EDTA, 0.1 % NaN$_3$, and 0.5 % triton X-100. For a monomeric solution, the protocols described by Teplow[93] and Hou et al.[94] were used with the minor modifications. The peptide was dissolved in 20 mM NaOH, sonicated and passed through a 0.22 µM syringe filter. A diluted protein solution (150 µM) in Tris-buffer (pH 7.2) and H$_2$O/D$_2$O ratios of 10, 25, 30, or 50 % was seeded with preformed sonicated fibrils (12 generations of seeding) and incubated at room temperature with agitation for one week. Growth and quality of the fibrils was monitored by EM of small aliquots. Representative pictures of fibrils, grown upon agitation and with and without 50 mM NaCl, are shown in Figure 2.5.

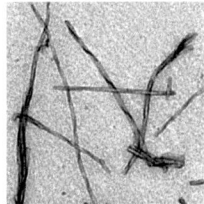 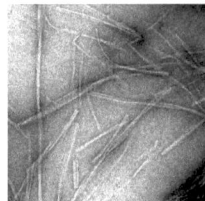

Figure 2.5. Electron microscopic image of $A\beta^{1-40}$ fibrils, with and without 50 mM NaCl. All solutions were agitated upon fibrillization.

Cu-doping of fibrils was achieved using Cu(edta) concentrations of 75 or 100 mM. Packing of the fibrils into rotors of a diameter of 3.2 or 4 mm was performed as described earlier for SH3 microcrystals. A table of the parameters that were optimized in the course of the thesis is given in Chapter 5.

3 PARAMAGNETIC RELAXATION

Paramagnetic compounds as transition metal ions can be used in protein NMR to selectively modify a number of parameters related to the NMR experiment.[95] The energetic dependence of the orientation of a paramagnetic compound on the direction of the external field can be employed for an alignment of proteins for RDC (residual dipolar couplings) determination.[96] In addition to the interactions between nuclei via dipolar and *J*-couplings, the nuclei of the protein can interact with the compounds' unpaired electrons if these are sufficiently close. Abundance of paramagnets in the protein under investigation can thus lead to modified chemical shifts and relaxation, providing additional information for resonance assignment and structure calculation.[80,97-99]

If the chemical shift without paramagnetic influence is known, the paramagnetically induced shifts (hyperfine shifts) can be mapped in order to obtain spatial information on the location of protein residues in respect to the paramagnet.[100,101] As agents that induce large contact shifts in combination with tolerable relaxation (so called *shift reagents*), mostly lanthanides(III) (except Gd) are used. Contact shifts δ^{pc} can be quantified using the expression:[102]

$$\delta^{pc} = \Delta\chi_{ax} \frac{(3\cos^2\theta - 1)}{r^3} + \frac{3}{2}\Delta\chi_{rh} \frac{\sin^2\theta \cos\varphi}{r^3}, \qquad (61)$$

where χ_{ax} and χ_{rh} are axial and rhombic components of the magnetic susceptibility tensor anisotropy, *r* is the metal-nucleus distance, and θ and φ are the polar angles between the metal-nucleus-vector and the principle axis of the susceptibility tensor.[103]

Similarly, longitudinal and transverse relaxation rates R_1 and R_2 are affected by the presence of unpaired electrons.[104,105] The spin relaxation effect arises from random fluctuations of the coupling energy between the spin and the magnetic field or other dipoles. Spin relaxation times τ_s are on the order of ns to ps. For short τ_s, high resolution NMR can still be performed, whereas larger τ_s values lead to fast nuclear relaxation. This effect is even more pronounced for a large total spin angular momentum $S = \Sigma(m_s)$. Table 3.1 gives τ_s and S for some paramagnetic metals in comparison to organic paramagnets.[106]

Paramagnetic Relaxation

Species	S	τ_s (s)
Organic compounds	1/2	$10^{-6} - 10^{-8}$
Cu(II)	1/2	10^{-9}
Cr(III)	3/2	$5 \cdot 10^{-9} - 5 \cdot 10^{-10}$
Mn(II)	5/2	10^{-8}
Gd(III)	7/2	$10^{-8} - 10^{-9}$
Ce(III)	5/2	10^{-13}
Er(III)	15/2	$8 \cdot 10^{-13} - 3 \cdot 10^{-13}$

Table 3.1. Spin angular momentum S and electron spin relaxation times τ_s for a selection of compounds, values taken from Banci et al.[106]

For *relaxation reagents*, contact shifts are mostly neglectable for distances larger than a few bond lengths due to a large energy difference between electronic ground state and excited states and an accordingly small magnetic anisotropy. Typical cases are Gd^{III}, Cu^{II}, Mn^{II}, Cr^{III} etc. Equation (56) (vide ultra) can be used for a description of the dipolar coupling between nuclei and electron. The resulting effect on the longitudinal relaxation of respective nuclei can be described by:[78,104,107]

$$\Delta R_1 = k/r^6, \qquad (62)$$

$$k = \frac{2\gamma_I^2 g_e^2 \mu_B^2}{15} S(S+1) \left(\frac{\mu_0}{4\pi}\right)^2 \left[\frac{\tau_c}{1+(\omega_I-\omega_S)^2 \tau_c^2} + \frac{3\tau_c}{1+\omega_I^2 \tau_c^2} + \frac{6\tau_c}{1+(\omega_I+\omega_S)^2 \tau_c^2}\right] \qquad (63)$$

with $\quad 1/\tau_c = 1/\tau_S + 1/\tau_I + 1/\tau_M \qquad (64)$

where γ_I is the nuclear gyromagnetic ratio of the nucleus, g_e is the free electron g-factor, S the quantum number associated with the electron spin, r is the electron-nucleus distance, ω_I and ω_S are the Larmor frequencies of the nucleus and electron, respectively. μ_0 and μ_B are the vacuum permeability and Bohr's magneton, respectively. τ_c is the effective correlation time, τ_S the electronic correlation time. While in solution, τ_I refers to the rotational correlation time of the molecule, in the solid state, τ_I represents the correlation time of local dynamics. τ_M denotes the correlation time of chemical exchange in solution.

Paramagnetic Relaxation

An additional source for relaxation is Curie relaxation, which goes back to an interaction of the nuclear spin with the induced magnetic moment of the electrons. For longitudinal relaxation, e.g.,[108]

$$R_1^{Cur} = \frac{2}{5}\left(\frac{\mu_0}{4\pi}\right)^2 \frac{\omega_I^2 g_e^4 \mu_B^4}{(3kT)^2 r^6} S^2(S+1)^2 \frac{3\tau_c}{1+\omega_I^2\tau_c^2}, \quad \text{with} \qquad (65)$$

$$1/\tau_c(Cur) = 1/\tau_M + 1/\tau_I \qquad (66)$$

Although being a severe limitation in solution state for increasing magnetic fields, Curie relaxation should be largely absent in the solid state, except for a small contribution which is caused by local mobility.[103,108]

3.1 Paramagnetic Relaxation Enhancement (PRE)

(Work in this chapter was published in publication #1 on page 183)

The high value for the electronic gyromagnetic ratio γ_e and the great set of electronic correlation times make paramagnetic relaxation a tuneable and potentially versatile effect. Relaxation reagents used as *spin labels* have been used for site specific peak attenuation.[97] This methodology, which is known from solution NMR, has also been applied to solid state NMR.[98] In the opposite case, where the paramagnetic species is freely dissolved in the protein solution, an overall PRE can be induced. With appropriate paramagnetic probes, this has been shown to allow for a largely shortened experimental time due to a faster return of the spins to their Boltzmann magnetization by increased R_1 rates.[77,79,109] By such, R_1, being more sensitive to the paramagnetic influence, approaches R_2, which stays mostly unaffected.[107]

For such a tailored PRE, transition metals have been shown to provide appropriate electronic correlation times (vide ultra). In contrast to organic paramagnets, specific binding of the metal ion is a potential drawback, which can be circumvented by addition of chelating compounds like edta[81], DTPA[110], DO2A[77] etc.

The applicability of paramagnetic doping in solid state NMR experiments has been shown by Ishii et al.[81] A described acceleration of data acquisition up to a factor of 10, however, cannot be used in traditional solid state NMR due to high power ^{1}H-decoupling during acquisition. Deleterious heating effects for high power decoupling in the course of a shortened duty cycle include sample degradation and probe damage. Temperature increase was quantified for decoupling at various decoupling strengths for two aqueous samples, containing 0 and 100 mM NaCl, respectively. The experiments employed 30 ms of decoupling and a recycle delay as indicated in Figure 3.1. The observed temperature increase ranges to 20 K in the absence of salt and a recycle delay of 300 ms. In contrast, for the same recycle delay, the solution easily reaches temperatures beyond 100 °C when 100 mM salt is present. Temperatures were calibrated by measurement of the chemical shift of the water resonance frequency in the range between 0 and 60 °C and extrapolated to higher chemical shift values. The resulting broad shapes of the resonance signal indicate a non-uniform distribution of temperature over the acquisition time and the rotor volume.[111] Thus, the data points in Figure 3.1 give a rough estimation about the order of rf-induced heating.

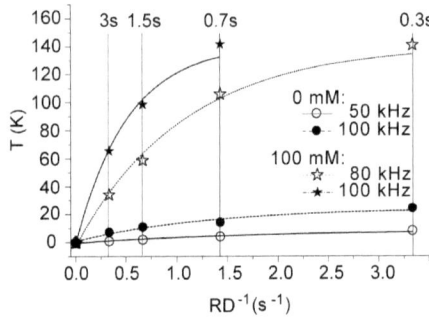

Figure 3.1: Increase of temperature upon decoupling. While moderate temperature increase occurs in the absence of salts, the increase with 100 mM salt present is intolerable for proteins using short recycle delays. The recycle delay is depicted in the upper row. Stars and circles refer to an amount of 100 mM and 0 mM NaCl with the decoupling strength indicated in the Figure. Acquisition took place directly after the decoupling interval.

The limitation of a severe temperature increase in the sample does not exist if paramagnetic doping is used for solid state NMR that is independent of high power decoupling. Furthermore, for deuterated systems, T_1 relaxation is particularly slow due to largely reduced dipolar relaxation by protons. In non-doped samples of 10 % proton-backsubstitution, T_1 typically amounts to 4.5 s. Use of high concentrations of paramagnetic agent can be used for a reduction of T_1 basically to T_2, resulting in vanishing recycle delays. This factor of 15 in measurement time upon PRE of 250 mM Cu(edta) largely facilitates proton detection by providing the possibility of experiments that are too insensitive otherwise (vide infra).

Figure 3.2 depicts inversion recovery data recorded for different concentrations of Cu(edta).

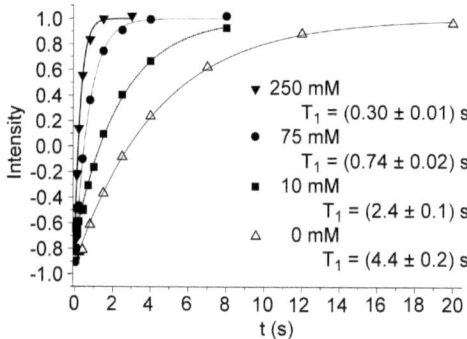

Figure 3.2: Inversion recovery upon doping with different Cu(edta) concentrations. Intensities depict the integral amide proton bulk signal in the first slice of an HSQC experiment modified by an initial inversion pulse. A reduction of T_1 by a factor up to ~15 was achieved by addition of Cu(edta) upon microcrystallization.

For a wide-spread use in solid state NMR, specific binding of the chelate to the protein is not desirable. In order to check for protein-chelate interaction, 2D HSQC spectra of doped and non-doped samples were recorded. Evidently, no chemical shift differences are observed, excluding the possibility of specific chelate-protein interactions. A representative region is displayed in Figure 3.3 A.

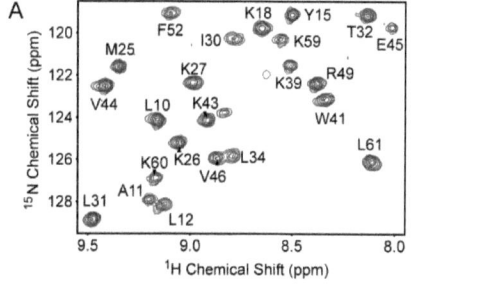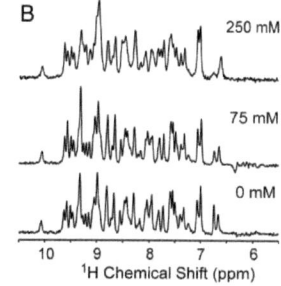

Figure 3.3: Representation of the influence of Cu-doping on chemical shifts (**A**) and linewidths (**B**). The spectra in A were recorded using an HSQC experiment and 0 (black) and 75 mM Cu(edta) doping (red), respectively. A depicts a representative region of the spectra. In order to represent signal attenuation effects, first proton 1D slices of an HSQC experiment were recorded in the presence of 0, 75, and 250 mM

Paramagnetic Relaxation

Cu(edta). The spectra were recorded at 700 MHz proton Larmor frequency using 24 kHz MAS and an effective temperature of 22 °C.

An accurate determination of chemical shift differences is displayed in Figure 3.4 B. The overall chemical shift difference upon doping with 75 mM Cu(edta) was calculated from ^{1}H and ^{15}N shifts as $\Delta\delta = 8.7 \pm 7.1$ Hz using $\Delta\delta = ([\Delta\delta(^1H)]^2 + [\Delta\delta(^{15}N)]^2)^{1/2}$.

Although T_1 is effected more strongly by paramagnetic influence, also T_2 decreases, resulting in a signal broadening for the effected spins. This feature occurs for doping with high concentrations of paramagnetic agents. However, up to moderate doping levels, only slight broadening effects were observed. The average proton linewidths amounted to 24.3 ± 5.5 and 31.5 ± 7.5 Hz for doping with 75 and 250 mM Cu(edta), respectively, in comparison to 24.6 ± 5.3 Hz without Cu(edta). A representation of ^{1}H linewidth increase is given in Figure 3.3 B, depicting first 1D slices of HSQC experiments for different doping levels. ^{15}N linewidth amount to 15.0, 16.0, and 19.9 Hz for a representative residue (L31) doped with 0, 75, and 250 mM Cu(edta). Site resolved proton linewidths are shown in Figure 3.4 A.

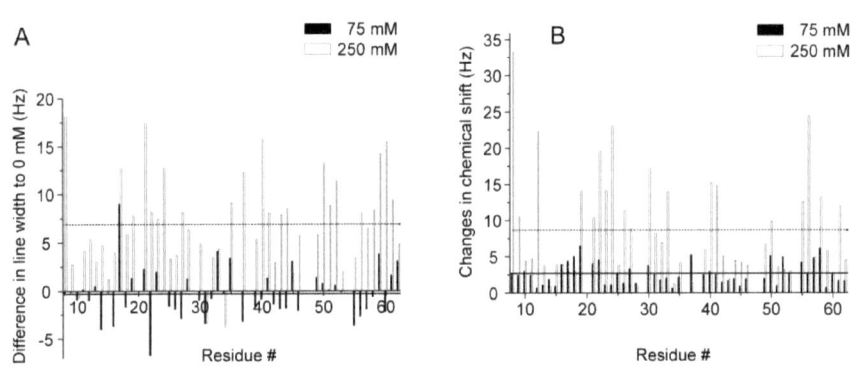

Figure 3.4. A) ^{1}H difference linewidths obtained for a 75 mM (closed rectangles) and 250 mM (open rectangles) Cu(edta) doped SH3 sample, compared to a copper-free reference sample as a function of the amino acid sequence. Horizontal solid and dashed lines in the Figure indicate the average value of the linewidth difference for samples containing 75 mM and 250 mM Cu(edta), respectively. B) Chemical shift differences of the samples containing 75 mM (closed rectangles) and 250 mM [Cu(edta)]$^{2-}$ (open rectangles) with respect to a sample containing no copper. Horizontal solid and dashed lines in the Figure indicate the

Paramagnetic Relaxation

average value of the chemical shift difference for samples containing 75 mM and 250 mM Cu(edta). The experimental conditions resemble those given in Figure 3.3.

3.2 Surface accessibility in the solid state

(Work in this chapter was published in publication #4 on page 183)

As denoted above, Paramagnetic Relaxation Enhancement is dependent on the distance r between the paramagnetic probe and the nucleus in a way proportional to r^{-6}. This gives rise to a site specific PRE being dependent on the minimal approachability of the paramagnet to the respective nucleus.

As mentioned earlier, spin labels have been used for site specific local PRE in order to access long range distance restraints in solution as well as in the solid state.[97,98] As an alternative to fixedly attached paramagnets, also dissolved relaxation agents have been used for a determination of solvent approachable sites in solution.[110,112,113] Although in the solid state, surface accessibility of proteins is largely determined by crystal-crystal contacts, solvent content on the order of 50 % still leads to vast contact surfaces between protein and bulk solvent. Numerous cases exist, where these interfaces are crucial for structure or function of the macromolecule.[114,115] Partially solvated structures have also been described in amyloid fibrils, providing important criteria for the understanding of the aggregation process.[116,117]

Although paramagnetic doping accounts for an overall increase of T_1 times, small differences for different residues can be mapped in order for a determination of nuclear approachability. Instead of using the amide bulk signal, residue resolved T_1 measurements were used. The pulse scheme used (see Figure 3.5 A) resulted from an HSQC by addition of an initial 180° ^{1}H inversion pulse. Respective raw data for the situation of 0 and 75 mM doping are plotted in Figure 3.5 B for selected amide H^N resonances.

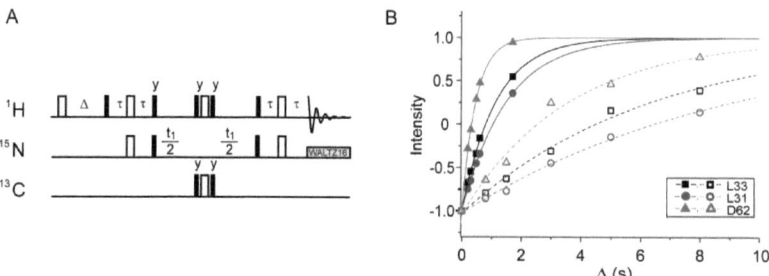

Paramagnetic Relaxation

Figure 3.5. A) Pulse sequence for the 2D inversion recovery experiments. Open bars refer to 180° pulses, closed ones to 90° pulses. τ was set to 2.4 ms, the recovery delay is depicted as Δ. Decoupling from ^{15}N-^{1}H-^{1}J-couplings during acquisition was performed using 2 kHz decoupling. **B)** depicts representative ^{1}H-T_1 curves for selected residues. The Figure displays the experimental data for the amide resonances of L33 (black), L31 (red), and D62 (green). Solid and dashed lines refer to crystals containing 75 mM Cu^{II} and 0 mM Cu^{II}.

The differences in T_1 of different residues along the backbone are present even without the influence of Cu^{II} and can be found for both, doped and non-doped SH3, in a consistent manner. Thus, these trends can be ascribed to factors inherent to the respective protein parts, like internal dynamics or chemical exchange processes. The trends (e. g. decreased T_1 times of the N- and C-terminus or in the flexible loop around P20) are in accordance with the dynamics of the protein (see Chapter 4.3). The existence and relative extent of the general T_1 site specificity is depicted in Figure 3.6, representing T_1 times with and without doping of 75 mM Cu(edta).

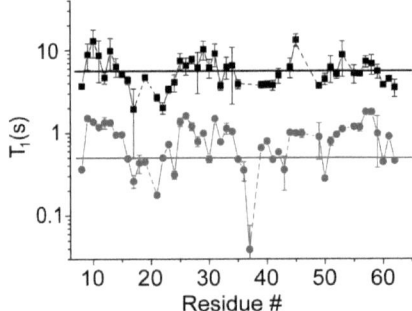

Figure 3.6. Longitudinal ^{1}H T_1 relaxation times for the SH3 domain of chicken α-spectrin in the absence of Cu^{II}-(edta) (black squares) and in the presence of 75 mM Cu^{II}(edta) (red circles). The average T_1 of all resolved amide protons amounts to 5.6 s and 0.49 s in the absence and presence of the paramagnetic solute, respectively.

Internal motion modulates longitudinal relaxation rates of different nuclei in close proximity to each other to a similar extent. This leads to a correlation between ^{15}N and ^{1}H longitudinal relaxation, both showing the same trends along the backbone. The consistency of relaxation trends for the different nuclei is present even though paramagnetic influence manipulates the intrinsic longitudinal relaxation rates, as can be seen by a correlation of ^{15}N longitudinal relaxation for a sample without doping (measured by Chevelkov et al.)[118] and ^{1}H longitudinal relaxation in the presence of

75 mM Cu^{II}. Consequently, the paramagnetic influence determines the apparent relaxation characteristics much less effectively than internal motion does. The respective correlation is depicted in Figure 3.7.

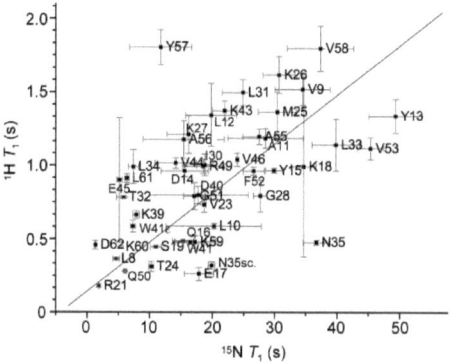

Figure 3.7. Correlation between ^{15}N longitudinal relaxation times T_1 in the absence and ^{1}H T_1 in the presence of Cu^{II}. In comparison to the role that local dynamics play for site-specific trends of relaxation, the presence of Cu^{II} has a severely weaker effect. Proton T_1 times were acquired according to Figure 3.5, the ^{15}N T_1 times were determined from measurements under the same conditions at 600 MHz Larmor frequency.

In order to eliminate relaxation effects from local flexibility etc., difference rates were extracted from micro-crystals with different levels of Cu(edta)-doping. Figure 3.8 shows difference rates extracted from three samples with 250 mM, 75 mM, and 0 mM Cu(edta). The site specific trend for relaxation enhancement is consistent for all three samples. The extent of site specific PRE is largely determined by the minimal distance r_0 between a specific site and the paramagnetic probe. In some parts of the protein, high R_2 rates due to large flexibilities and accordingly low signal to noise can be found. For the following steps, only difference rates accounting for ΔR_1 (75 mM – 0 mM) were used.

Paramagnetic Relaxation

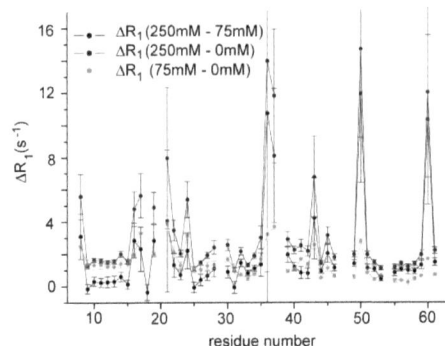

Figure 3.8. Difference relaxation rates extracted from three protein preparations containing 250 mM, 75 mM, and 0 mM Cu^{II}. Similar trends for the difference H^N R_1 relaxation rates ΔR_1 for all three combinations of the three data sets hint to a consistent site-specific PRE effect by Cu(edta) doping.

In contrast to the situation of two fixed spins, the paramagnet can diffuse freely in the space excluded by the protein, leading to additional sites where the Cu-chelate induces relaxation. The additional positions with distances $r > r_0$ contribute less to the observed PRE each, however, their number increases with r^3. The effective PRE resulting from the whole space beyond the protein surface can be determined by an integration over potential residence sites.[119,120] This can be done e. g. by an integration over a half sphere with angles φ, θ and radius r':

$$\Delta R_{1,\mathit{eff}} = k \cdot r_{\mathit{eff}}^{-6}, \qquad \text{with} \qquad (67)$$

$$r_{\mathit{eff}}^{-6} = \int_0^{r'_{\max}} \int_0^{2\pi} \int_0^{\pi/2} r^{-6} \cdot r'^2 \sin\theta \, d\theta \, d\varphi \, dr'. \qquad (68)$$

Thereby, the distance r between an H^N proton and each site in space can be determined using

$$r = \sqrt{r_0^2 + r'^2 + 2r_0 r' \cos\theta}. \qquad (69)$$

The function can be modified for consideration of convex and concave surface shapes by choosing the upper limit for the integrand θ larger or smaller than 90°, respectively. Figure 3.9 gives a pictorial representation of the half-spherical integration of solvent accessible space.

Paramagnetic Relaxation

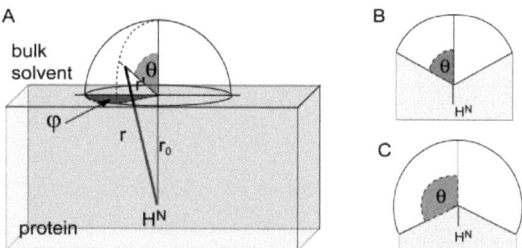

Figure 3.9. A) Representation of a half spherical integration over solvent accessible space beyond the protein-solvent surface. To a good approximation, the effective PRE can be obtained using a cut-off value of r'max = 6 Å. **B)** θ can be integrated up to values smaller or larger than $\pi/2$ in order to reflect concave or convex curvature of the surface.

A more accurate consideration of locally specific surface shapes has been proposed by Pintacuda and Otting.[110] A grid search for positions in which the ligand does not collide with the protein provides detailed information on the accessible space accounting for the integration. However, a crystal structure as a prerequisite would exclude many potential applications.

Numerical evaluation of the integral can be provided by standard mathematics software. Logically, to a good approximation, the effective PRE has a r_0^3 dependence in respect to the minimal distance r_0. Figure 3.10 depicts the result of a numerical integration for each of the scenarios, plane, convex, and concave curvature, in comparison to simple, appropriately scaled r_0^{-3} functions (represented as crosses).

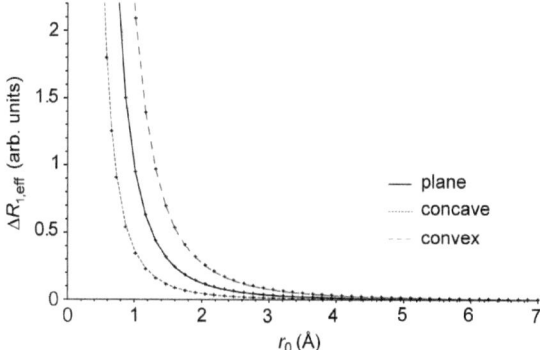

Figure 3.10. Representation of the numerical integration (lines) in comparison with accordingly scaled r_0^{-3} functions (crosses). Besides a plane surface (center), the plot shows concave and convex surface curvatures (dotted lines) using $\theta_{max} = 60°$ and $120°$, respectively.

The effective PRE value, which is dependent on the space considered for the integration, converges in respect to the cut-off r'_{max} for $r'_{max} \gg r_0$. Nevertheless, even larger relative errors for small r_0 are small in comparison to PRE of easily approachable sites, since the respective PRE values are small. Thus, integration up to 6 Å is sufficient for a reasonable accuracy. This can be seen by numerical evaluation of the integral in dependence of r'_{max} for different r_0 values. Figure 3.11 depicts the relative accuracy of the integration by the help of a factor $Q(r'_{max})$ using

$$Q(r'_{max}) = \frac{1}{r_{eff}^6(r'_{max})} \cdot \left[\lim_{r'_{max} \to \infty} \left(\frac{1}{r_{eff}^6} \right) \right]^{-1}. \tag{70}$$

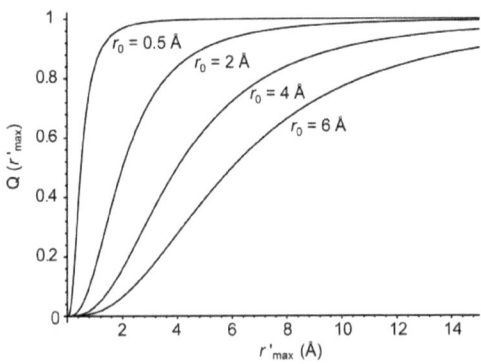

Figure 3.11. Convergence of the spherical integral for different r_0, represented by the factor $Q(r'_{max})$. For amide protons located directly beneath the surface, small cut-off values r'_{max} are sufficient, while an effective distance for amide protons further away can only be calculated employing sufficiently large spheres. However, the absolute contribution to the difference rate ΔR_1 for amide protons with large r_0 is rather small. The poor convergence for those protons does therefore not compromise the analysis. Normalization of Q in respect to the true PRE was performed for a better visualization of their steepness as the curves for larger r_0 have tiny PRE values.

In order to correlate the T_1 relaxation rates induced by paramagnetic doping with the H^N-to-surface distances accounting for different protein residues, a Connolly surface calculation[121,122] was used.

Paramagnetic Relaxation

This employed a 15-mer of the SH3 X-ray structure (pdb-code 2NUZ)[60] and a probe radius of 4 Å.[123] By a stepwise determination of H^N atoms up to a distance to the Connolly surface increasing by ~0.5 Å, each proton was assigned to a distance r_0.

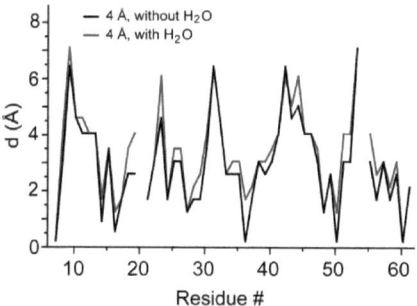

Figure 3.12. Distances r_0 from H^N protons to the protein-solvent interface determined by a Connolly surface calculation dependent on the residue. A maximum distance of ~7 Å was found for the SH3 domain.

Figure 3.13 depicts a model of a Cu(edta) complex. The approximate radius of 4 Å does not reflect the minimal distance from chelate surface to the unpaired electron. This can rather be assumed ~1 Å considering a closest nucleus-surface distance of ~2.5 Å and a Cu van-der-Waals radius of 1.5 Å.[124]

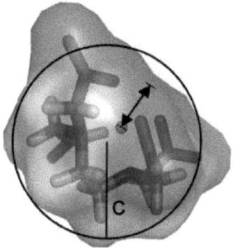

Figure 3.13. For the exploration of the chelate accessible volume in the crystal structure, Cu(edta) was assumed to be a perfect sphere with radius c = 4 Å. The minimal distance between chelate surface and Cu nucleus (arrow) is approximately 2.5 Å.

A correlation between the differential R_1 rates upon doping with 75 mM Cu(edta) and the distance r_0 between H^N proton and protein-solvent interface is depicted in Figure 3.14. The fit of the data by

Paramagnetic Relaxation

a r_0^{-3} function reveals a horizontal offset of 0.5 Å, which is in accordance to the expected additional distance from chelate surface to the Cu^{II} unpaired electron. More interestingly, a vertical offset of 0.7 s^{-1} hints to a uniform T_1 increase. This site-unspecific PRE is due to the dipolar based averaging of T_1 relaxation predominantly by proton-proton spin diffusion. Although dipolar couplings are strongly suppressed by deuteration, the residual amount of protons in the sample still leads to a certain degree of spin diffusion. This is beneficial since PRE is most often used for an accelerated data acquisition rather than site-specific surface accessibility determination.

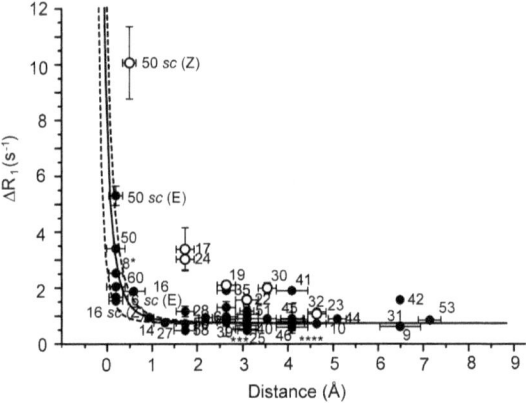

Figure 3.14. Difference relaxation rates ΔR_1 versus distance to the chelate accessible surface for all residues. Open circles reflect residues which are close to side chain hydroxyl groups. All other residues are represented in black. Dotted lines refer to concave (θ_{max} = 60°, lower curve) and convex surface shape (θ_{max} = 120°, upper curve), respectively.

The outliers in the correlation 3.14 (depicted with open circles) hint to an interesting side effect of PRE. The HN protons of these residues are in close proximity to vicinal hydroxyl groups. These are thought to quickly exchange with the bulk solvent,[125] which is in close proximity to the paramagnetic agent. Spin diffusion of exchanged OH-protons with backbone amides results in a strongly enhanced effective PRE (relay-PRE) for respective sites. The selection of amides in close proximity to OH (see Figure 3.15 A) thus coincides with the awkward values of the residues depicted as open circles in Figure 3.14. A general dependence on the proximity of these processes according to the 6th potency makes the interaction at vicinal sites even more significant in comparison to those with large distances to OH-groups.

Electrostatic attraction to positively charged or polar side chains (Arg-, Lys-side chains) as a potential source for increased PRE can be excluded, since no amide protons have a significant proximity to those species. This is depicted in Figure 3.15 B and C.

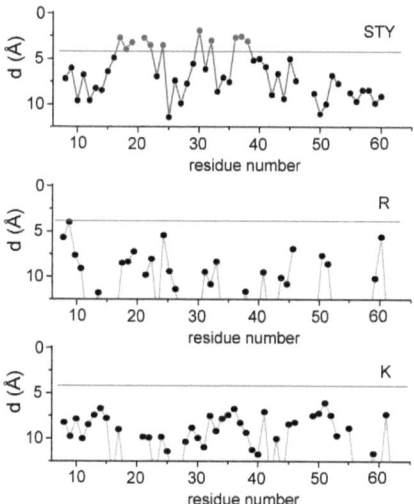

Figure 3.15. Distances between H^N and hydroxyl groups (top), arginine side chains (center), and lysine side chain NH_3^+ (bottom), respectively (for distances < 15 Å). The red dots and lines mark the H^N of close proximity to respective species employing a cut-off of 3.5 Å.

Assuming the existence of OH-relayed paramagnetic relaxation, also water might function as a mediator for PRE. This effect besides direct PRE might also be responsible for part of the vertical offset of the fit in the Figure 3.14 if a certain "breathing motion" of the protein leads to influx of water molecules to sites that are inaccessible for direct PRE. By such, L8 fits to the curve better if the Connolly probe radius for distance determination is set to the water radius. This might also explain the unexpectedly high PRE values of W41, W42, and N35, which do not appear to have significant proximity to hydroxyl-groups.

As can be seen in Figure 3.15, I30 is exceptionally close to a hydroxyl group, namely T37-OH (d = 1.89 Å). Nevertheless, only a moderate PRE can be observed for this amide H^N. This observation is expected, since a rapid exchange with bulk water is strongly reduced due to an H-bond between T37-OH and N35-C**O**NH. Figure 3.16 depicts the local geometry of this site as seen from the X-ray structure.

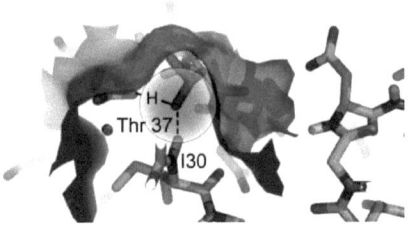

Figure 3.16. SH3 structure around residue T37. Due to localisation in the interface between two symmetry related molecules and involvement in an H-bond with N35, the exchange of the OH group of T37 with bulk water protons is likely to be restricted. Consequently, no strong PRE is observable for I30, although the intermolecular distance of I30-H^N to T37-**O**H is exceptionally short (1.89 Å).

4 METHODOLOGICAL IMPROVEMENT OF ^{1}H DETECTED SOLID STATE NMR

4.1 The proton content

While traditional solid state NMR relies on the chemical shift information of heteronuclei, this work is focussed on the inclusion of ^{1}H into the group of spectroscopically observed nuclei. Fast MAS has provided the possibility of proton detection in the direct acquisition dimension, as proton linewidths scale linearly with the rotor period.[126] As mentioned in the Introduction, protein ^{1}H linewidths are on the order of 400 Hz at 40 kHz spinning and 750 MHz.[55] In combination with a high sensitivity, this has provided 3D experiments with a reasonable resolution.

A decrease of dipolar couplings due to a reduced proton content has been shown to result in significantly reduced linewidths and an accordingly enhanced resolution. Linewidths are on the order of 140 Hz and 40 Hz for ^{1}H and ^{15}N, respectively, if only exchangeable sites are protonated.[58] Furthermore, sensitivity is accordingly enhanced in comparison to fully protonated samples.

For an even further reduced linewidth, extensive deuteration has been combined with partial back-substitution of exchangeable protons. Achievable linewidths of around 20 and 10 Hz in the ^{1}H and ^{15}N dimension, respectively, are comparable to the linewidths found in solution NMR.[59] This approach is useful for site-resolved investigation of local structural and dynamic investigation of biomolecules. Spin diffusion, which would otherwise average signal intensities and relaxation characteristics of a specific site, is greatly reduced. This has been shown to be beneficial e. g. for accurate determination of backbone motion in the solid state,[127,128] or detection of water molecules in the micro-crystalline lattice.[72]

Figure 4.1 represents the protonation level of these 3 scenarios (fully protonated, deuterated with 100 % back-exchange, and with 10% back-exchange) for the SH3 domain in a pictorial way.

Methodological work on the SH3 domain

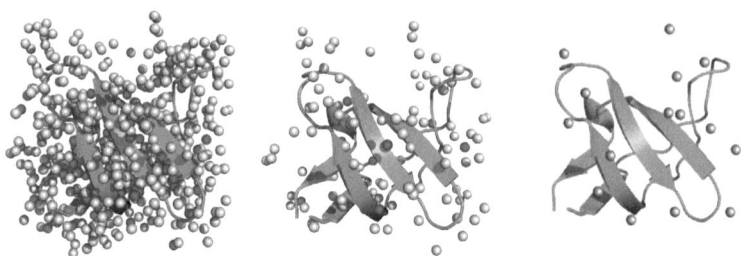

Figure 4.1. Pictorial representation of the protonation content in a uniformly protonated (left), uniformly deuterated sample with 100 % (center) and 10 % proton back-exchange at exchangeable sites.

While further proton substitution yields sharper lines and longer FIDs by reduced dipolar couplings, sensitivity is compromised by a decrease of NMR-active species. Therefore, a specific ^{1}H-level seems appropriate dependent on the application.

For a single pulse excitation, an increased protonation level in principle scales the peak integral. However, the signal to noise achievable for samples of different protonation differs for pulse schemes including additional magnetization transfer steps. Cross Polarization on the one hand is a fast transfer and provides little loss of magnetization due to relaxation. INEPT transfer steps on the other hand, relying on the lengthy evolution of small J-couplings, suffer from significant relaxation for protonation levels higher than 30–40 %, referring to a MAS frequency of 24 kHz. For little magnetization loss due to relaxation, however, INEPT can in principle provide quantitative transfer, whereas CP yields a maximum of around 70 % transfer efficiency. Furthermore, CP transfer efficiency is severely decreased in case of protein mobility. This effect dues to changing Hartmann-Hahn conditions upon motion. (A detailed comparison of magnetization transfer methods for different flexibility is given in Chapter 4.3)

In contrast to traditional NMR of fully protonated samples, duty cycles can be almost arbitrarily short, since no high-power ^{1}H decoupling has to be applied. This renders the longitudinal relaxation time T_1 an additional factor for the signal intensity, since a reduced T_1 enables a faster scan repetition. Signal to noise scales with the square root of the number of applied scans.

A comparison of line broadening, T_1 and T_2 times by increasing protonation from 10 to 100 % in exchangeable sites for a ν_{MAS} = 24 kHz is depicted in Figure 4.2. The integral does not represent real peak intensities.

Methodological work on the SH3 domain

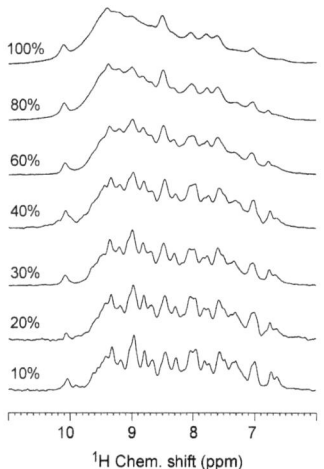

^{1}H/^{2}D	^{1}H T_1 (s)	^{1}H T_2 (ms)	^{15}N T_2 (ms)
100 %	0.68	11.4	9.6
80 %	0.76	23.2	13.6
60 %	0.85	29.8	23.6
40 %	1.67	33.1	37.5
30 %	2.13	51.4	38.4
20 %	2.76	55.3	57.3
10 %	4.28	60.6	75.8

Figure 4.2. Left: Amide bulk signal of the SH3 domain for various levels of ^{1}H back-substitution at exchangeable sites (H$_2$O/D$_2$O ratios are denoted on the left). The curves were scaled to equal height. The depicted spectra were recorded with a CP back-and-forth transfer to ^{15}N for water suppression. Right: Relaxation times dependent on the protonation level of the exchangeable sites, both according to Akbey et al.[129]

Line broadening was observed not to be equal for each protein site. While many residues were completely attenuated in an H/N-correlation at >60 % back-substitution, certain residues could still be observed at the highest protonation degree. This behaviour is most likely due to internal dynamics, which are responsible for effective reduction of dipolar ^{1}H-^{1}H couplings at flexible sites. This effect has been reported even for fully protonated samples before.[130] According H/N-correlation spectra as well as sections through two representative resonances for rigid and mobile protein parts are depicted in Figure 4.3.[129]

Methodological work on the SH3 domain

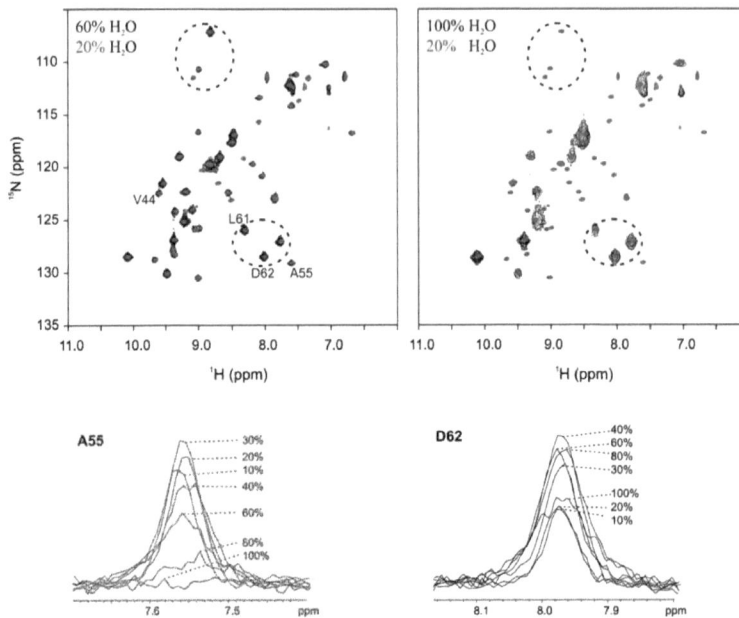

Figure 4.3. Top: H/N correlation spectra recorded for different protonation degrees. Most residues are broadened severely going from 60 to 100 % ^{1}H. However, certain resonances are retained due to motion of the protein site. Bottom: Exemplary resonances of a stiff (A55) and a flexible residue (D62). D62 shows much higher relative signal intensities for high protonation degrees than A55.[129]

For most residues, a protonation level of around 30 % turns out to the best compromize between slow relaxation and a high abundance of NMR-active nuclei. Special consideration for *J*-based triple resonance experiments will be found below.

4.2 Solution state like 3D experiments

(Work in this chapter was published in publications #2 and 6 on page 183)

Detection of ^{1}H chemical shift provides access to an additional source of structural and dynamic information. Signals of the respective groups can be separated more easily by the increased dispersion due to an additional dimension in multidimensional experiments. Thus, inclusion of ^{1}H chemical shifts tends to enable unambiguous assignment of overlapping resonances.

The exclusion of ^{1}H chemical shifts from data acquisition due to dipolar broadening is one of the main differences between traditional solid state NMR and solution NMR. ^{1}H-detected experiments are performed in solution state as the proton's gyromagnetic ratio is largest among the usually abundant nuclei of biomolecular NMR, see Chapter 1. This fact is usually taken into account by using ^{1}H as a starting point for magnetization transfer as well as for the acquisition dimension. For deuterated proteins, which are protonated only at the exchangeable sites, so called out-and-back experiments are used to satisfy the requirement of ^{1}H detection and ^{1}H start magnetization for optimal signal to noise.[19]

Parallel to solution NMR experiments, INEPT transfers can be used in the solid state if homo- and heteronuclear dipolar couplings are reduced to a level where relaxation is not detrimentally fast. Besides a quantitative transfer efficiency in the absence of relaxation, INEPT transfer exclusively refers to chemically bonded nuclei. Thus, in contrast to dipolar based polarization transfer, clean correlations without non-specific artefact signals due to spatially close nuclei can be obtained.

3D experiments for backbone assignment

The HNCO experiment[19,131] is the most straight-forward version of a series of multidimensional ^{1}H-detected experiments for backbone chemical shift correlation. The large ^{15}N-^{13}CO coupling of ~15 Hz and a magnetization transfer to only one ^{13}C coupling partner makes it the most sensitive triple resonance out-and-back experiment. Transfer from solution state to solid state implies omission of gradient pulses, a simplified solvent suppression, and optimization of ^{1}H decoupling. Figure 4.4 shows the HNCO pulse sequence with and without use of heteronuclear decoupling. We used a constant-time evolution of ^{15}N within the N/C-INEPT in order to provide optimal signal to noise. All transfer steps were chosen to base on INEPTs (in contrast to HMQC versions using multiple quantum coherences).

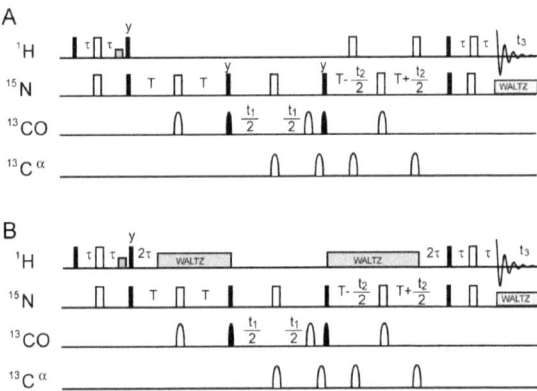

Figure 4.4. Pulse scheme for an HNCO experiment in the solid state, using no heteronuclear decoupling (**A**) or Walz-16[132] as a decoupling sequence for ^{1}H during transverse ^{15}N magnetization (**B**).

Signal to noise amounted to approximately 15 % of a corresponding HSQC experiment. This makes it possible to run a full 3D HNCO within some hours of experimental time. For solid state NMR with partial proton back-substitution, theoretical signal to noise of ~35 % compared to an HSQC should be obtainable, assuming 100 ms ^{15}N T_2.[128] This discrepancy might be due to poor RF inhomogeneity during pulses on the one hand and a reduction of transverse relaxation times T_2 of all involved nuclei due to the presence of paramagnetic Cu^{II}. The constant-time evolution for ^{15}N allows for t_2^{max} < 23 ms. Although this results in a truncation of the signal in the ^{15}N dimension ($T_2(^{15}$N) being ~ 30 ms), the achievable resolution of a 3D was found to be largely sufficient for a maximal ^{15}N incrementation within the ^{15}N/^{13}C INEPT step. At the same time, the sensitivity is not compromised by additional relaxation during the ^{15}N evolution. Use of two symmetric pulses for refocusing of the $^1J_{N/CO}$-coupling does not cause erroneous ^{15}N chemical shift evolution during the (non-negligible) pulse duration.

Selective ^{13}C pulses were applied using soft rectangular pulses for on-resonance CO excitation in order to reduce artefacts. G3 Gaussian shape pulses[133] were used for off-resonance ^{13}C$^{\alpha}$ excitation. Bloch-Siegert phase shift during the duration of the shaped pulse was compensated by an additional G3 pulse after on-resonance inversion of ^{13}C. Values for T and τ were chosen 12 and 2.3 ms. We used Waltz-16[132] for decoupling for both, ^{1}H and ^{15}N (pulse scheme B). For the latter case, 2 kHz decoupling power was used. ^{1}H decoupling can in principle be continued during the ^{13}C evolution period. This did not turn out to change signal to noise (in the limit of error) due to the absence of larger *J*-couplings between ^{1}H and ^{13}C. Furthermore, incidentally created conditions for recoupling of ^{1}H,^{13}C dipolar coupling could be avoided renouncing ^{1}H decoupling at this point. Since decoupling focuses on *J*-couplings, we only used up to 5 kHz for ^{1}H decoupling. This also takes in account the lengthy periods of around 37 ms in each scan over which decoupling takes place. Before start of ^{1}H decoupling, antiphase H_zN_y coherence was initially refocused and created over a period 2τ after and before the ^{1}H,^{15}N INEPT, respectively. Signal to noise was very sensitive to misadjustments between the decoupling power and the Waltz 90° pulse length and needed extraordinarily careful optimization. Furthermore, except for an improved solvent suppression, no significant gain in sensitivity was observed in comparison to pulse scheme A for a 10 % protonated sample. Water suppression, however, was almost negligible for SH3 due to a good crystal packing upon ultra-centrifugation while packing the rotors. For this reason, ^{1}H decoupling was omitted for SH3 samples of 10% back-protonation. Figure 4.5 depicts a 1D version of the experiment using 128 scans and the pulse schemes A and B of Figure 4.4, respectively.

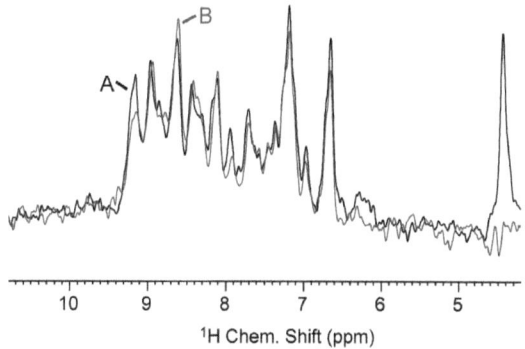

Figure 4.5. Comparison between pulse sequences A (black spectrum) and B (red spectrum) of Figure 4.4. Signal intensities of the bulk amide signal after 128 scans do not display systematic differences except for solvent suppression quality. Procession was performed using exponential multiplication of 25 Hz.

Figure 4.6 depicts a 2D version of an HNCO experiment recorded with pulse sequence A modified by a continuous ^{1}H decoupling also during transverse ^{13}C magnetization. The indirect dimension was incremented up to 14.5 ms in order to allow for a maximum signal to noise.

Methodological work on the SH3 domain

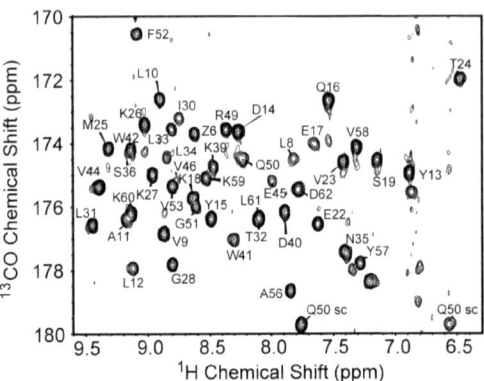

Figure 4.6. The first slice of an HNCO experiment in the solid state. $t_1(^{13}CO)$ was incremented up to 14.5 ms. Exponential apodization of 5 Hz and gaussian multiplication (-10 Hz, shift of the bell by 0.1) was used for the direct and indirect dimension, respectively. The excellent resolution due to sharp lines is even increased upon incrementation of the 3rd dimension (^{15}N).

Sequential assignment via CO chemical shifts

Resonance assignment is the requirement for most subsequent NMR data evaluation. One important part of this process is the protein backbone assignment, since a consequent linkage via chemical bonds allows for small regular steps of a sequential assignment. For isotopically enriched proteins, this is usually done via experiments correlating backbone resonances to other either intra- or interresidual backbone nuclei in close vicinity. Typical experiments for sequential assignment of deuterated proteins in solution refer to CO or C^α and C^β chemical shifts in combination with the amide moiety.[19,131,134-138]

The solution state HNCACO experiment provides correlation between the amide moiety and the intraresidual ^{13}CO chemical shift as the main correlation.[139,140] Due to the approximate similarity of ^1J and ^2J (~11 Hz and 7 Hz, respectively)[138], a second correlation to the interresidual C^α with slightly less intensity is observed. Figure 4.7 represents the magnetization transfer pathway in comparison to an HNCO experiment.

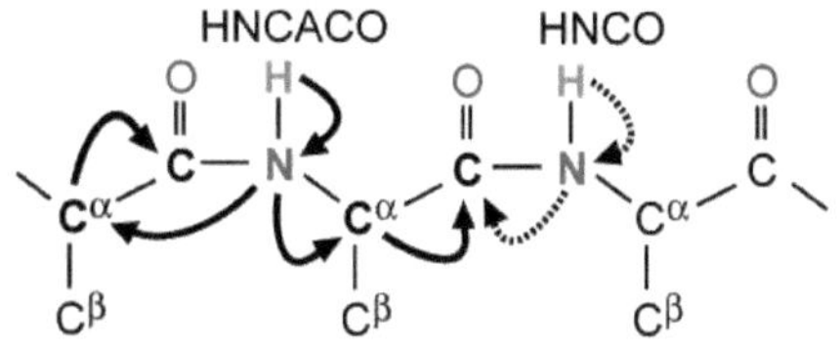

Methodological work on the SH3 domain

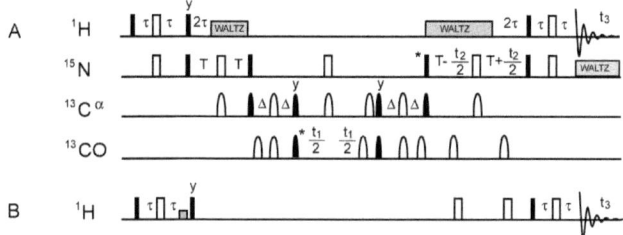

Figure 4.7. Magnetization transfer pathways for the HNCACO and the HNCO. HNCACO (left) transfers ^{15}N magnetization to both CO_i and CO_{i-1}. HNCO (right) gives correlations only for the interresidual CO_{i-1}. The Figure shows only the "out"-transfer of ^1H magnetization, which is followed by an equivalent transfer back. Sequential assignment is straight forward by combination of HNCACO and HNCO.

In order to implement this solution state experiment for solid state NMR, the respective solution state sequence was used and modified by omission of gradient pulses, substitution of water gate water suppression by a simple purge pulse and optimization of ^1H decoupling during transverse magnetization on heteronuclei.

Figure 4.8 depicts the modifications of the pulse sequence used for deuterated solid samples.

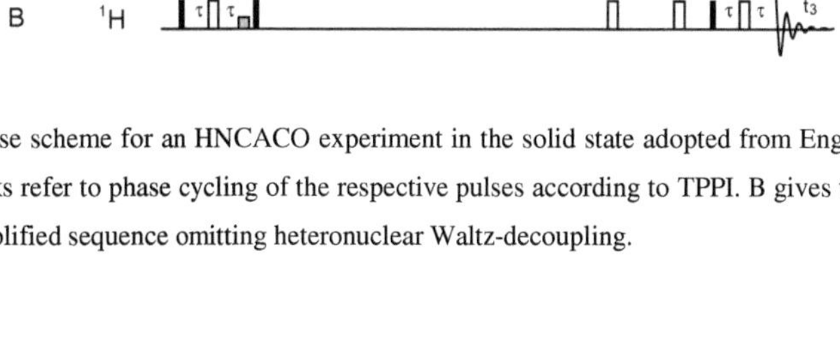

Figure 4.8. Pulse scheme for an HNCACO experiment in the solid state adopted from Engelke and Rüterjans.[140] Asterisks refer to phase cycling of the respective pulses according to TPPI. B gives the modified ^1H pulses of a simplified sequence omitting heteronuclear Waltz-decoupling.

For a reduction of artefacts, all on-resonance pulses were applied as soft rectangular pulses, trimmed such that off-resonance nuclei were unaffected. This required the carrier frequency on the ^{13}C channel to be changed from C^α to CO frequency and back in between the 90° pulses of the C^α/CO INEPTs. Due to the reasons stated above, we stuck to the sequence without decoupling in the course of SH3 experiments of low degrees of back-protonation. Comparing 1D versions of HNCACO and HNCO correlations (see Figure 4.9), an approximate four fold loss in signal to noise occurs due to pulse imperfections of the additional pulses and the relay step for full evolution of the $^1J(C^\alpha\text{-CO})$ coupling. This makes up 16 ms of additional transverse $^{13}C^\alpha$ coherence, in which evolution of passive C^α-C^β couplings of a magnitude almost as large as the active C^α-CO coupling reduce observable magnetization. Loss of signal to noise due to this step can in the future be largely reduced if deuterium decoupling and a four channel probe are used and if C^α-C^β couplings are refocused by appropriate selective pulses on $^{13}C^\beta$.

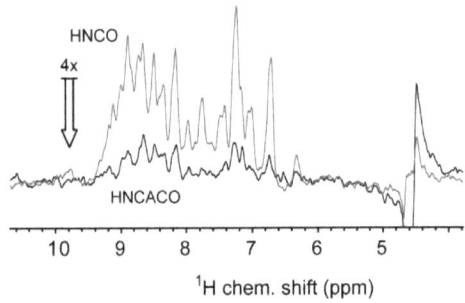

Figure 4.9. Intensity comparison between HNCACO and HNCO recorded under identical conditions. The additional relay step and an increasing number of pulses in the sequence is assisted by a four-fold decrease in signal to noise. Data was acquired using 128 scans, 3.5 kHz Walz-16 ^{1}H decoupling during transverse ^{13}C periods and 40 ms acquisition at 24 kHz MAS, 700 MHz ^{1}H Larmor frequency and an effective temperature of 22 °C. Data were processed using 20 Hz exponential line broadening. Artefacts at ~4.8 ppm result from incompletely suppressed water signal.

Figure 4.10 shows a representative excerpt of the sequential walk based on ^{13}CO chemical shifts for the SH3 domain. The signal to noise for the experiment amounted to approximately 25:1 for the intraresidual correlation. This value corresponds to three days of experimental time, using 20 ms maximum t_1 incrementation and use of a 10 % ^{1}H back-substituted, 150 mM Cu^{II}-doped sample of SH3 with 530 ms recycle delay.

Interresidual correlations rely on the slightly less intense ^2J-coupling of $^{15}N_i$ and $^{13}C^\alpha_{i\text{-}1}$ (~7 Hz instead of 11 Hz for the intraresidual ^1J-coupling). Consequently, interresidual cross-peaks were observed to yield ½ to 2/3 of the intensity of the intra-residual peak.

Methodological work on the SH3 domain

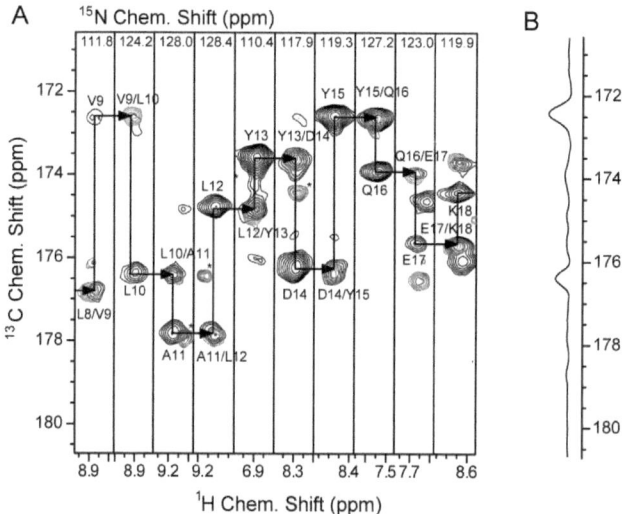

Figure 4.10. A) Sequential walk through ^{13}CO resonances of the SH3 domain of chicken α-spectrin. HNCACO correlations are depicted in black, while those of a corresponding HNCO are shown in red. The excerpt shows the resonances between residues L8 and K18. Except from the flexible regions along the N-terminus and the distal loop, similar data quality was achieved for the rest of the protein resonances. **B)** Representative column (taken from strip Y15/D14) for a representation of the signal to noise ratio between intra- and interresidual correlations.

For the SH3 domain, CO based sequential assignment experiments yielded a nearly complete backbone assignment. Problematic residues yielding low signal to noise are located in regions of high internal mobility undergoing slow motion. The N-terminal residues E3 to E7 and C-terminal D62 did not yield resonances above the noise level in the HNCACO. In respect to the HNCO, this applies to N-terminal residues E3, K6, E7, and D62. Further residues of insufficient signal intensity are located in the flexible distal loop around N47. This is the case for both experiments, although the HNCO also resolves E45, V46, and R49, which do not yield enough signal to noise in the HNCACO.

In comparison to ^{13}C assignment via ^{13}C direct acquisition, homonuclear couplings do not compromise the accuracy of CO chemical shifts even in fully ^{13}C labeled material. This is due to the fact that a simple inversion pulse on C^α refocuses all homonuclear ^1J-couplings. This is in contrast

Methodological work on the SH3 domain

to direct acquisition of ^{13}C, where homonuclear decoupling during acquisition[141] or spin state selective experiments[142] have been proposed for refocusing of homonuclear J-couplings.

In order to estimate the achievable accuracy of CO based assignment using solution state sequences in the solid state, ^{13}C T_2 decay rates were determined. Figure 4.11 A shows a relaxation curve of CO transverse magnetization. Unlike ^{13}C direct or C$^\alpha$ indirect detection, the measured T_2 of ~40 ms is directly reflected by an according linewidth $\Delta_{HWFM} = T_2/\pi$, as depicted in Figure 4.11 B (bottom). Only field inhomogeneities or local motion induce a comparably small additional broadening effect. A comparison to the potential of a sequential assignment via C$^\alpha$ and C$^\beta$ resonances shows an approximate gain in accuracy of a factor of 5, taking into account that duplets arising from $^1J(C^\alpha/C^\beta)$-couplings are usually not resolved in order to reduce signal to noise losses. This results in an apparent linewidth of ~100 Hz. Use of 4-channel probes providing ^{2}H-decoupling can in principle reduce this value. If a constant time evolution period is used,[137] truncation can be reduced, however, the additional relaxation would lead to a reduced signal to noise in comparison to real-time evolution with refocused J-couplings.

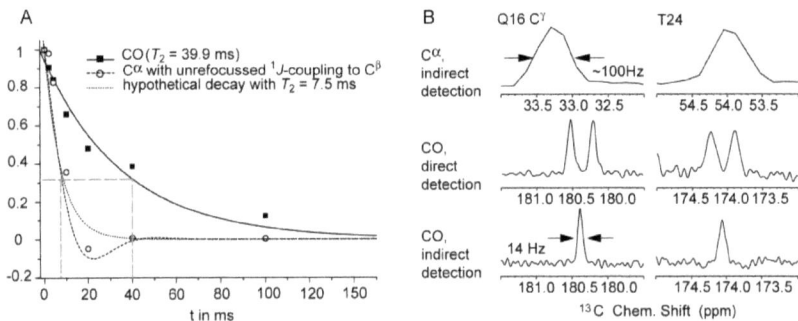

Figure 4.11. Reflection of the T_2 decay at 600 MHz in the linewidths of ^{13}C direct and indirect acquisition, comparing ^{13}C$^\alpha$ and ^{13}CO based experiments. **A)** T_2 decay for CO and C$^\alpha$ with refocusing of $^1J(C^\alpha/CO)$ and $^1J(C^\alpha/^2H^\alpha)$ but without refocusing of $^1J(C^\alpha/C^\beta)$. **B)** For indirect CO detection, assignment experiments can yield high accuracy. This cannot easily be achieved by experiments based on C$^\alpha$ chemical shifts or direct ^{13}C detection. The slices of CO indirect detection are taken from the ^{1}H/^{13}CO 2D in Figure 4.12.

Linewidths of 14–18 Hz give rise to hardly any signal overlap even in only one dimension. Besides being a highly resolved source for other biochemically relevant information like relaxation values or paramagnetic shifts and attenuation e. g., this makes a sequential assignment merely based on CO chemical shifts possible. The resolution in CO chemical shifts for the SH3 domain is represented by Figure 4.12. The spectrum was recorded with a t_1^{max} of 100 ms and processed without use of apodization. Use of apodization and linear prediction may be used to even further decrease residual shift degeneration. Marked are the resonances that were used for the withdrawal of the sections in Figure 4.11 B. Using the CO based approach in addition to C^α based sequential assignment can lead to unambiguous assignments also for larger proteins, where either of the methods would result in ambiguous chemical shifts.

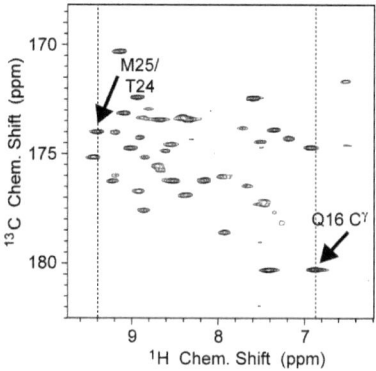

Figure 4.12. Representation of the resolution that CO indirect detection can yield at 600 MHz even without use of apodization. Nearly completely unambiguous assignments are possible with the mere use of CO chemical shifts alone and can be supported by use C^α based assignments to give unambiguous assignments even for larger proteins. Dashed lines depict the position of the slices shown in Figure 4.11 B.

In order to evaluate to what extent the good resolution obtained for CO chemical shifts rivals solution state NMR, relaxation rates of dissolved proteins were estimated by using the CSA induced relaxation as the bases for a minimal values of R_2. Although ^{13}CO relaxation originates from other sources as well like CSA/dipole-dipole cross-correlation between ^{13}CO and $^{13}C^\alpha$ and between ^{13}CO and ^{15}N, at field strengths larger than 400 MHz for protons, ^{13}CO relaxation is largely dominated by this contribution. In contrast to dipolar interactions, which are neglectable and independent on the field, and cross-correlation contributions scaling linearly with the field, the CSA influence scales quadratically with the magnetic field.[143] The respective contribution to the transverse relaxation rate $R_{2,CSA}$ is given as[143]

Methodological work on the SH3 domain

$$R_{2,CSA} = (d_{CSA}^2 \omega_C^2 / 6)[4J_{C'}(0) + 3J_{C'}(\omega_C)],\tag{71}$$

with a generalized CSA coupling constant $d^2_{CSA} = (1/3)\Delta\delta^2[1 + \eta^2/3] = 6.2 \cdot 10^{-9}$,[144] where $\Delta\delta^2$ and η represent the anisotropy and the asymmetry of the CSA tensor, respectively. $J_{C'}(\omega)$ is the spectral density function at frequency ω.

Minimal relaxation rates of ^{13}CO in solution (exclusive consideration of $R_{2,CSA}$) in dependence of the molecular correlation time τ_C are shown in Figure 4.13.

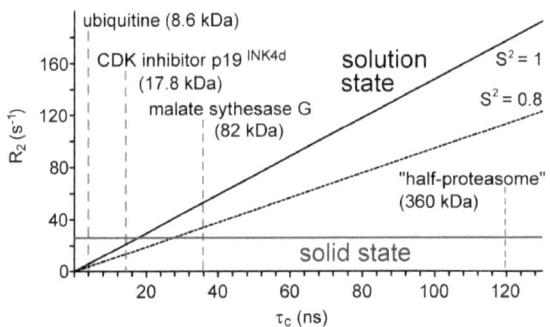

Figure 4.13. Comparison of carbonyl transverse relaxation rates R_2 in solution (black) and in the solid state (red) at 14.1 T (600 MHZ ^{1}H Larmor frequency). The dashed black line shows ^{13}CO relaxation rates in solution assuming a generalized order parameter S^2 of 0.8, while the solid black line is drawn upon assumption of an order parameter of 1. Whereas R_2 in solution strongly depends on the correlation time τ_c, the lower limit for the linewidth in solid-state NMR (depicted for micro-crystalline SH3, red) is molecular weight independent. Vertical, dashed lines indicate typical values for the motional correlation time τ_c for several proteins in solution.

The Figure depicts relaxation rates in solution at 14.1 T (corresponding to a ^{1}H Larmor frequency of 600 MHz) for two different order parameters S^2 using

$$J(\omega) = \frac{2}{5}\left[\frac{S^2 \tau_C}{1+(\omega\tau_C)^2} + \frac{(1-S^2)\tau_i}{1+(\omega\tau_i)^2}\right],\tag{72}$$

according to the model-free approach of Lipari and Szabo.[145]

Values of 1 (no internal motion) and 0.8, representing the order of motion of most proteins (between 1 and 0.6) were chosen for the S^2.[144] For a better comprehension, solution-state correlation times were provided with representative proteins of a respective molecular size. Shown are the correlation times of ubiquitin, the CDK inhibitor p19^{INK4d}, malate synthase G and the "half proteasome", having a molecular size of 8.6, 17.8, 82, and 360 kDa.[146-149] For intermediate or large proteins, the plot shows that carbonyl resonance linewidth in the solid-state can potentially be smaller than the linewidth obtainable in solution. We observe a break-even between linewidth in the solid state and in solution for motional correlation times τ_c of ~ 25 ns, corresponding to a molecular size on the order of 30 kDa for globular proteins. Further reduction of dipolar relaxation in the solid state (by increasing MAS frequencies, e. g.) might push the margin at which solid state lines become sharper than in solution even further down to smaller proteins.

Sequential assignment via C^α and C^β chemical shifts

Sequential backbone assignment in the solution state does usually not rely on carbonyl chemical shifts alone. Although the accuracy would be sufficient for an unambiguous assignment for small proteins, a complementary approach is based on the chemical shifts of C^α and C^β.[131,136] For protonated proteins, "straight-through" experiments are usually more sensitive than the according "out-and-back" experiments used for deuterated proteins. These are not feasible in our case due to missing H^α protonation. While the sensitivity of an HNCACB[136] is significantly lower than the HNCA experiment,[19,131,138] the assignment process can be facilitated substantially by using two resonances for a sequential correlation. Similar to the HNCO experiment (see above), magnetization is transferred from ^1H to $^{13}C^\alpha$ via $^1H/^{15}N$- and $^{15}N/^{13}C^\alpha$-INEPTs. The peculiarity of the J-based magnetization transfer from N^H to C^α is the approximate similarity of ^1J- and ^2J-couplings to inter- and intra-residual C^α (^1J and ^2J of ~11 and ~7 Hz, respectively), as already denoted above. This results in a bidirectional pathway to both $^{13}C^\alpha$ nuclei. Although giving rise to sequential correlations within one experiment, sensitivities naturally go down in comparison to respective experiments that yield only one correlation.

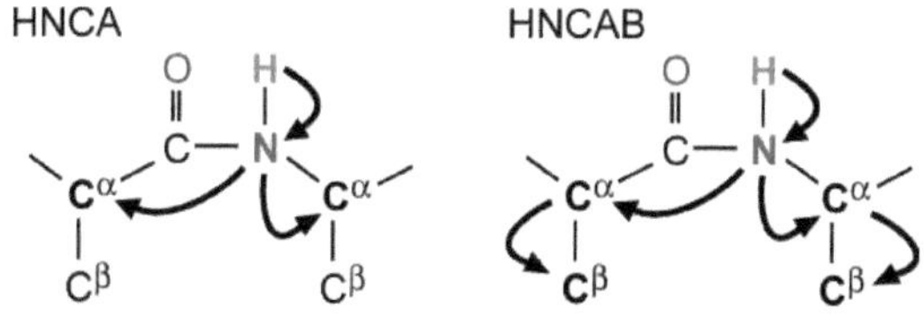

Figure 4.14. Pictorial presentation of the magnetization transfer pathway of an HNCA (left) and an HNCACB experiment (right). Only half of the pathway of the out-and-back experiments is depicted. In any

Methodological work on the SH3 domain

case a bidirectional transfer to C^α_i and C^α_{i-1} is observed due to comparable ^1J and ^2J for N-C^α. In the HNCACB, an additional relay step in the pulse sequence gives C^β correlations in addition to C^α using only one ^{13}C dimension.

For the generation of a solid state HNCA experiment, the HNCO experiment described above was adapted by a shift of the carrier frequency to C^α (and respective changes in the soft pulse offsets). Incrementation in the indirect dimension was performed such that the spectral range covered 45 ppm and t_{1max} amounted to 9 ms. Further acquisition in the $^{13}C^\alpha$ without use of specific refocusing pulses on C^β would lead to a decrease in intensity due to the oscillation of the respective coherence (N_zC_y) with $\cos(\pi J^{C\alpha C\beta} t)$, which means a zero-crossing at approximately 14 ms. The fact that both, inter- and intraresidual correlations show up in the spectra allows for a sequential "walk along the backbone" and a respective assignment of backbone resonances. Figure 4.15 displays an excerpt of the sequential walk along the backbone based on C^α chemical shifts.

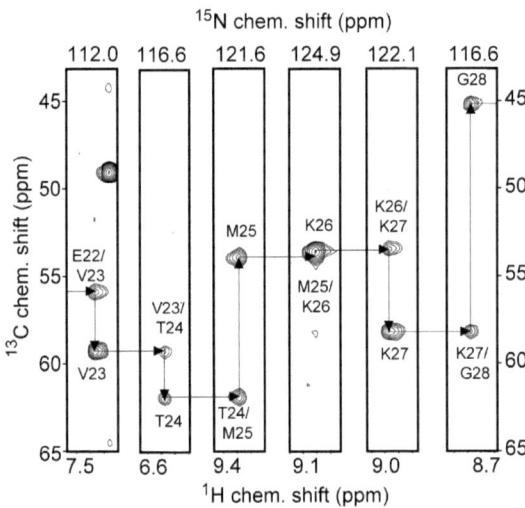

Figure 4.15. Representative part of the sequential walk along the backbone based on C^α chemical shifts. The according 3D spectrum was recorded on a 4.0 mm sample with a MAS frequency of 14 kHz within 3.5 d. A protonation level of 10 % was used for the exchangeable sites in combination with a PRE of 150 mM $[Cu^{II}(edta)]^{2-}$.

Methodological work on the SH3 domain

The depicted spectrum was recorded using a 150 mM Cu^{II} doped sample in a 4 mm rotor at 14 kHz spinning frequency. The assignment of the SH3 resonances using this approach turned out to be almost complete using 10 % backsubstitution of exchangeable deuterons against protons. The N-terminal residues up to E7 and residues in the flexible distal loop around N47 were not observed. Inherently low signal to noise of flexible residues resulted in visible resonances above noise level partly only for a 3.2 mm sample with a back-protonation of 25 % and 75 mM Cu^{II} doping. Additional facilitation for these residues can be achieved by use of coherence selective sequences (see section 4.3).

An additional transfer of C^α magnetization to the C^β has been performed in the solution state in order for a correlation of both respective resonances to the amide group ($^1H^N$ and ^{15}N chemical shifts).[136] This can be performed using a relay step with evolution of only half of the $C^\alpha C^\beta$ anti-phase operator and provides the occurrence of all four C^α and C^βs adjacent to the amide group. Although this results in an elegant experiment with a large content of information, especially for a low overall sensitivity, an experiment for exclusive correlation to either one would be appreciable. A full evolution of the C^α-C^β couplings over 14.3 ms, however, interferes with both, concomitant transverse relaxation and the evolution of ^1J-couplings to the directly bonded ^2H, which could not be decoupled with the present setup. We used a relay of 7 ms as a compromize. All other features of the experiment were taken over from the ones described before. A representation of the pulse sequence is shown in Figure 4.16.

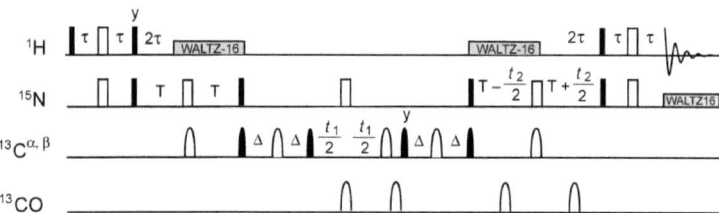

Figure 4.16. Pulse sequence for the HNCACB experiment. Experiments were performed in an out-and-back manner. ^1H decoupling was applied after refocusing the H,N antiphase coherence of the initial INEPT using the WALTZ-16 scheme[132] with a power of approximately 7 kHz. In the HNCO/HNCA experiment τ, T, and Δ were set to 2.3, 12.0, and 3.6 ms, respectively. On-resonance selective pulses on $^{13}C^{\alpha,\beta}$ were applied as

Methodological work on the SH3 domain

soft rectangular pulses, while ^{13}C off-resonance pulses on ^{13}CO were implemented as G3 shapes.[133] Decoupling during acquisition was achieved using WALTZ-16,[132] adjusting the ^{15}N RF field to 2 kHz.

For SH3, in addition to the depicted pulse programs also the minor modification of omitting the ^{1}H-decoupling sequence was used, employing composite pulses on the ^{1}H-channel simultaneous to the off-resonance soft pulses (as described for the HNCO and HNCACO), which did not change the performance of the experiment. Use of decoupling concomitantly during transverse ^{13}C magnetization turned out in a slightly reduced signal to noise (~90-95 %) probably due to resonance phenomena leading to partial reintroduction of dipolar interactions.

We used 70 ppm spectral width and (for the reasons stated above) maximal ^{13}C indirect acquisition times of 9 ms. For a 25 % back-substituted SH3-sample with PRE of 75 mM Cu^{II}, an HNCACB experiment could be recorded within 3 days, yielding an approximate signal to noise of 30:1. The relative intensity in comparison to an HSQC amounts to only 3 %. Besides an overall length of the sequence of ~70 ms and the application of 27 pulses in total, this is due to distribution of the H^N polarisation to four peaks of inter- and intraresidual C$^\alpha$ and C$^\beta$ correlation. Figure 4.17 depicts a section through the 3D at ^{15}N and ^{1}H chemical shifts of Tyr15 for representation of relative signal to noise for the contributions of different correlations to the spectrum.

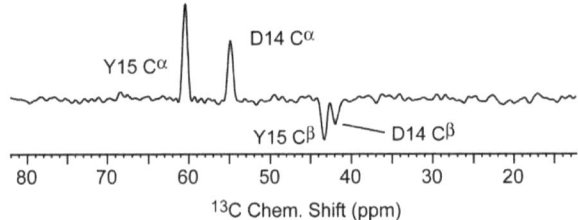

Figure 4.17. Representative slice taken out of the 3D at ^{1}H (8.48 ppm) and ^{15}N chemical shift (119.1 ppm) of Y15, recorded at an effective temperature of 22 °C. Signal to noise amounts to ~30:1 for a 25 % ^{1}H back-substituted 75 mM Cu^{II} doped SH3 3.2 mm sample with a restricted volume rotor within 3 d. Interresidual signals are reduced to ~60 % of the intraresidual correlation, C$^\beta$ correlations yield ~45 % of the intensity of C$^\alpha$ correlations.

By the splitting of resonances into three dimensions, each carbon strip can be assigned to HN resonances unambiguously, which facilitates the sequential walk along the protein primary sequence. Providing a two fold matching of connectivities, assignment is unambiguous for the SH3 domain

Methodological work on the SH3 domain

of α-spectrin even though line widths amount to around 100 Hz for ^{13}C due to homonuclear J-couplings and ~60Hz for ^{15}N due to the constant-time evolution. For a 4.0 mm sample 10 % protonated in exchangeable sites, doped with 150 mM Cu^{II}, signal to noise amounted to only 16:1 after 4.5 d. Although 12 and 5 residues miss one or two correlations, respectively, sequential correlations can still be withdrawn for most of the resonances. A strip plot from an experiment on a 25 % sample, recorded at a MAS frequency of 24 kHz, is depicted in Figure 4.18, showing sequential assignment for residues M25 to T32. This spectrum yields a signal to noise of ~30:1 after three days.

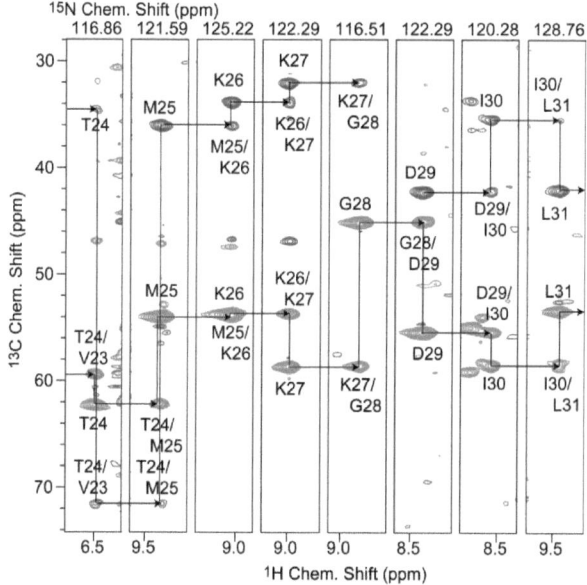

Figure 4.18. Walk along the backbone in an HNCACB experiment with a 25 % proton back-exchanged 3.2 mm sample with PRE of 75 mM Cu^{II}(edta). The Figure shows a representative part of the full sequential assignment. Adjacent residues can be found since a signal set usually consists of the correlations between H^N, N^H and C$^α_{i-1}$, C^α_i, C$^β_{i-1}$, and C^β_i. A signal to noise of ~30:1 was achieved after 3 d.

For a facilitated assignment of HNCA or HNCACB spectra, showing both, intra- and interresidual correlations of amide group and respective ^{13}C resonances, solution state uses HN(CO)CA or HN(CO)CACB experiments. These experiments yield interresidual correlations exclusively. They

Methodological work on the SH3 domain

can be performed using an additional relay step on ^{13}CO before magnetization is transferred further to C^α or C^α and C^β. In the solid state, experimental conditions have to be adjusted as shown in the previous cases for solution state triple resonance experiments. The pulse scheme for an HN(CO)CA is implicitly given in Figure 4.8, realized by an interchange of CO and C^α pulses (by a change of respective carrier frequencies). Figure 4.19 represents a demonstrative part of a 3D in combination with strips in a HNCA experiment.

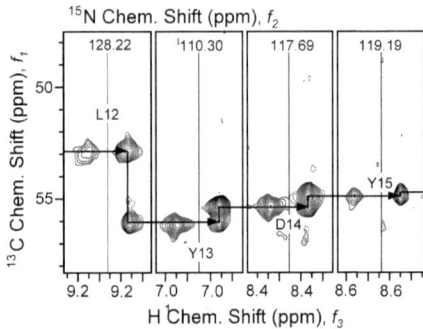

Figure 4.19. Representative part of the HN(CO)CA experiment correlating the amide group with the inter-residual C^α resonance only. The HN(CO)CA (red) is shown in combination with an HNCA experiment for sequential assignment, which yields both, inter- and intraresidual correlations, with the latter one being the more pronounced one.

In accordance to the HN(CA)CO experiment, the sensitivity of an HN(CO)CA experiment suffers from an additional magnetization transfer in comparison to HNCA or HNCO. In comparison to HN(CA)CO, however, only one correlation per residue is observed, and the relay nucleus (^{13}CO) does not evolve homonuclear (C^β) or heteronuclear J-coupling that are difficult to refocus. Although ^{13}CO CSA based relaxation is higher, the sensitivity of the HN(CO)CA was observed slightly better. We observed a signal to noise of the 3D of ~25:1 after 3 d of acquisition. In comparison to an HSQC of the same sample (yielding a signal to noise of ~30:1 in 15 min), this is a sensitivity of 4.5 %. We refer to a sample with 10 % back-substitution in amide sites and 150 mM Cu-PRE. Like for the HNCACB, use of 25-30 % back-substitution is assumed hence to increase the sensitivity by a factor of ~2.

Methodological work on the SH3 domain

CP-based triple resonance experiments

Especially for systems with increased relaxation due to higher protonation degrees or PRE, INEPT steps may lead to severe relaxation losses in comparison to CP based experiments. This can e. g. be the case for a higher degree of ^1H interactions, as described in Chapter 5. In this work, higher protonation levels were used in particular in the course of the Aß$^{1-40}$ assignment. Additionally, we observed this system to suffer from significantly higher relaxation rates than the SH3 domain even for identical proton concentrations, maybe due to dynamics or different behaviour towards Cu^{II}-PRE (see Chapter 3). In these cases, an exchange of INEPT magnetization transfer by Cross Polarization (CP) steps turned out to be a reasonable expansion of the solid state out-and-back triple resonance experiments. The details of this strategic change in respect to pulse schemes and resulting spectra can be found in Chapter 5.

4.3 Differential relaxation and spin state selection

(Work in this chapter was published in publication #9 on page 183)

In solution, an overall tumbling motion determines the values of the spectral density functions that are important for R_1 and R_2. In solid state NMR, however, the only motion encountered are local dynamics. The common measure for this kind of dynamics is the model-free approach, which was introduced by Lipari and Szabo[145] and extended by Gronenborn and coworkers.[150] Here, motion is described by a model-free order parameter S^2 and a correlation time τ for each regime of motion. Therein, the order parameter can be understood as how the resulting state of a dynamic process correlates with the initial state after a long time. If this motion is not completely random but rather restricted to a certain range of results (e. g. the direction of an interatomic bond vector), the final state will correlate much with the initial one and the parameter S^2 is larger than 0.

As described in Chapter 1.3, relaxation is coupled to motional processes in the protein. Using the model-free approach for different kinds of motion, the spectral density function is considered

$$J(\omega) = \frac{S^2 \tau_r}{1+\omega^2 \tau_r^2} + (1-S_f^2)\frac{\tau_f}{1+\omega^2 \tau_f^2} + S_f^2(1-S_s^2)\frac{\tau_s}{1+\omega^2 \tau_s^2}, \qquad (73)$$

with the indices s and f denoting the values for slow and fast motion. The first term describes the rotational diffusion, using the index r and being relevant only for molecules in solution. In the following, only a qualitative description of motion will be used, referring to the kind of motional regime (fast, slow) and the amplitude of the respective motion.

Methodological work on the SH3 domain

The different contributions to the observed overall relaxation, like e. g. dipolar or CSA-induced relaxation, only appear for certain regimes of motion, while for motion of a different time scale, they are largely absent. This is the case for very fast motion (ps) or very slow motion. Although – to a first order approximation – solid state proteins are immobilized and do not comprise the dynamics we find in solution or gas phase, local mobility is largely present also in the solid state. Besides the overall rotation introduced by MAS of the rotor, internal mobility of e. g. methyl groups[151], side chains[152] or loops[153] make up a whole set of sources relevant for relaxation. Backbone motion of the SH3 domain has been characterized by the measurement of ^{15}N T_1 relaxation times,[127] the ^{1}H-^{15}N dipole, ^{15}N CSA cross correlated relaxation rates $\eta^{DD/CSA}$,[118] and ^{1}H-^{15}N dipolar couplings[154]. The analysis revealed vast presence of internal motion in the solid state. Most prominently, the C-terminus (residue D62) was shown to undergo large amplitude motion in comparison to protein parts included in rigid β-sheet structures e. g..

An exemplary excerpt of the large amount of data needed for an accurate determination of all, order parameters S^2, amplitudes and time scales of different motional processes, is given in Figure 4.20, presenting ^{15}N R_1 relaxation. The represented values are influenced by ^{1}H/^{2}H exchange processes, since these reduce the acquirable signal in just the same way as longitudinal relaxation does.

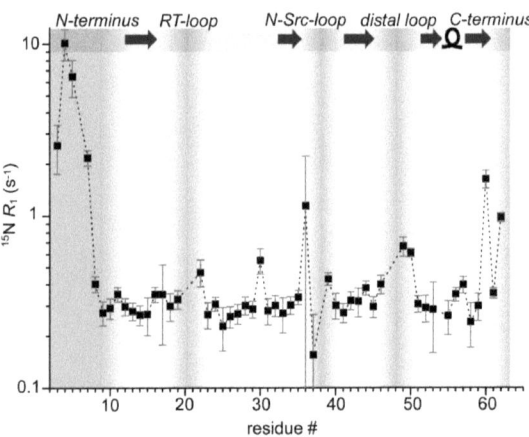

Figure 4.20. Longitudinal ^{15}N relaxation rates R_1 of a 25 % ^{1}H back-exchanged SH3 sample doped with 75 mM Cu^{II}(edta). Although PRE has site specific influence on R_1, the main contribution to longitudinal

relaxation can still be assumed to be due to dynamics (see Chapter 3). The data were recorded at 700 MHz, 24 kHz MAS and an effective temperature of 22 °C.

Residues undergoing slow motional processes are difficult to access in dipolar based experiments because the dipolar couplings used in the course of a Cross Polarisation are a function of time on the time-scale of the motion itself. This problem is hard to overcome with traditional Cross Polarization. Using scalar based transfers, however, residues undergoing slow motion are enabled to appear in correlation spectra. This can be seen in Figure 4.21, comparing two representative regions of the H/N correlation of the SH3 domain, recorded with CP and INEPT transfer, respectively. We find residues of flexible regions that have not been observable before using traditional approaches, proving that scalar transfer very much facilitates an observation of slow molecular motion.

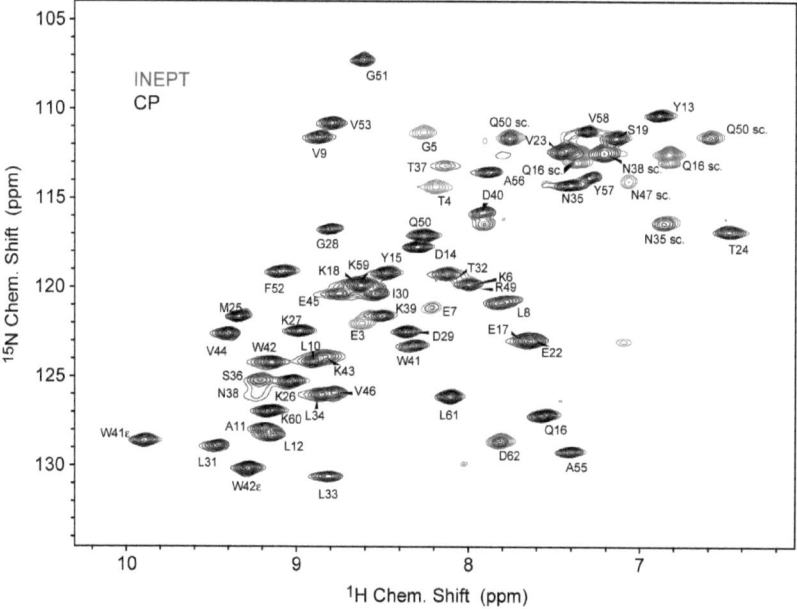

Figure 4.21. Comparison of dipolar and scalar magnetization transfer in an H/N correlation experiment, depicted in black and red, respectively. Flexible residues show up only in the INEPT-experiment. Both experiments were recorded and processed under the same conditions. The spectra were obtained at 600 MHz for a deuterated sample of the SH3 domain of α-spectrin, in which labile sites have a proton content of 25 %. The sample was spun at 24 kHz at 22 °C effective temperature.

Methodological work on the SH3 domain

The residues that show up in correlation spectra exclusively are weak and comparably broad in both, the proton and the ^{15}N dimension. While rigid residues show a linewidth of around 20 and 10 Hz in the ^{1}H and ^{15}N dimension, the additional resonances have linewidths of up to 50 and 30 Hz for ^{1}H and ^{15}N dimension. Figure 4.22 shows M25 as a largely immobile residue in comparison to mobile T4 and E7. The spectrum was recorded for a sample with 10 % proton content at exchangeable sites at 24 kHz MAS and an effective temperature of 22 °C. The additional broadening must be due to the difference in mobility and an according cross-correlated cross relaxation, since dipolar relaxation alone cannot be assumed to differ throughout the primary sequence.

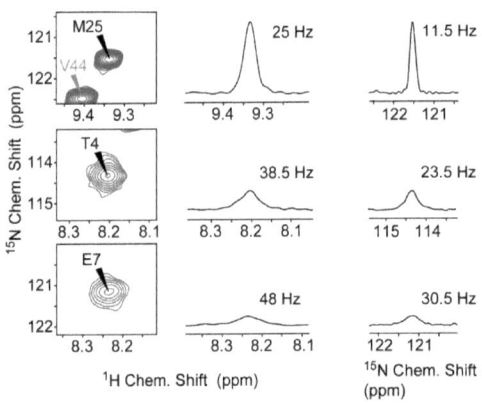

Figure 4.22. Line shapes of rigid residues and those undergoing slow motion, represented by M25 and by T4 and E7. The H/N-correlation is based on INEPT-transfers using the HSQC pulse scheme. The columns in the center and at the right are traces along the ^{1}H and ^{15}N axis, respectively. The spectra was recorded within 20 h with a t_{1max} of 100 ms. No apodization was used for procession. The sample was spun at 24 kHz at an effective temperature of 22 °C.

As mentioned in Chapter 1.3, relaxation interference results in mutual cancellation of dipolar and CSA induced relaxation for certain spin states. Thus, separation of spin states with convenient relaxation properties from those with inconvenient properties can yield long-lived NMR coherences. As shown by Pervushin et al.,[155] Transverse Relaxation Optimised Spectroscopy (TROSY) may select only one quad image out of the four possible coherences that are obtained with traditional methods. Selection of the coherences with a long lifetime results in an accordingly sharp linewidth and reduced relaxation loss in the course of transverse magnetization periods of the NMR experiment. Extraction of the narrow component can be achieved using a spin-state-selective magnetization transfer in combination with an appropriate phase cycle. In contrast to an INEPT transfer, an ST2-PT (Single Transition to Single Transition-Polarization Transfer) building block uses intermediate zero- and double-quantum coherence in order to yield observable H⁻ coherence. This avoids mixing of spin states.

Using the TROSY sequence, a pulse scheme for the measurement of spin-state dependent transverse relaxation rates of ^{15}N in an amide group can be created. This scheme is shown in Figure 4.23.

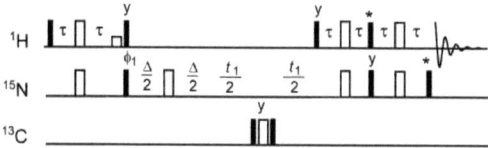

Figure 4.23. A) Pulse scheme for the spin state selective determination of transverse relaxation rates R_2 of the coherences N⁻H$^\alpha$ and N⁻H$^\beta$. Open and closed bars refer to 180 and 90° pulses, the bar with a reduced height is a 1 ms water purge pulse. For phase sensitive indirect acquisition, φ_1 was incremented according to TPPI. Inversion of pulses marked with an asterisk allows for swapping of the selected spin state. (Since only ^{15}N R_2 is measured, the first marked pulse should be inverted here.)

According to the TROSY scheme, one quad-image is selected by a simple two-step phase cycle, in which the first 90° pulse as well as the receiver are cycled in parallel 90° steps. By inversion of the last proton 90° pulse, the phase cycle selects the opposite ^{15}N coherence. Inversion of the last ^{15}N 90° pulse gives evolution of the H⁻ coherence under influence of the opposite ^{15}N spin state.

Figure 4.24 shows the ^{15}N transverse relaxation rates of N⁻H$^\alpha$ and N⁻H$^\beta$ coherence for a sample of 25 % proton amount in exchangeable sites at 24 kHz MAS with an effective temperature of 22 °C at 700 MHz ^{1}H Larmor frequency. A) and B) refer to 3 °C and 22°C effective temperature. C) gives a representation of the SH3 structure, with loop regions plotted in brown, β-sheet-regions plotted in blue and the short α-helix shown in orange.

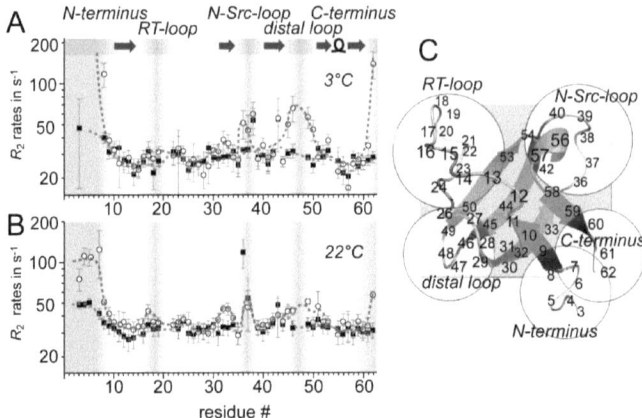

Figure 4.24. Relaxation rates for coherences N'H$^\alpha$ (narrow component) and N'H$^\beta$ (broad) component at 700MHz. Data were recorded for two different temperatures, 3 °C (**A**) and 22 °C (**B**). Transverse relaxation rates R_2 of sharp and broad component are depicted as black squares and white circles, respectively. Dashed lines are drawn in order to guide the eye. Secondary structure is indicated in the upper part of the Figure. Loop regions are shaded in brown. The protein internal motion and the effect of differential relaxation are dependent on the temperature as well as on the specific protein site. Especially high values for differential relaxation are found in the C-terminal D62, the N-terminus and the distal loop involving residues 45 to 50. Rates are depicted on a logarithmic scale. At low temperature, relaxation rates were not measurable for the N-terminal residues due to insufficient signal to noise. **C)** Pictorial representation of the loop regions in a structural model of SH3.[60] The flexible N-terminus (up to K6) is not refined in the crystal structure and shown with an arbitrary conformation.

Relaxation rates of N'H$^\alpha$ and N'H$^\beta$ are almost identical for the rigid parts of the molecule. This is true for large parts of the β-sheet regions as well as for the short α-helix between residues 54 and 58. Only minor parts of the β-sheet between residues 22 and 35 display differential relaxation. On the other hand, loop regions seem to undergo motion on the slow motion time scale. Particularly the N- and C-terminus display large differences between N'H$^\alpha$ and N'H$^\beta$ coherence (with a difference of ~50 Hz at 22 °C). A difference of ~70 Hz can be observed for E7, e. g.. For the RT and distal loop, the difference amounts to approximately 10-15 Hz. Residues 47 and 48 are expected to yield even higher differential rates, however, they have not yet been assigned due to low signal to noise due to their dynamics. In addition to an effective temperature of 22 °C, data were measured at 3 °C. At this temperature, the N-terminal residues do not give enough signal intensity for reliable relaxation data. The other loop regions display an enhanced differential relaxation. Even the outer parts of the distal loop (E45, V46, R49, and Q50 give rise to differential relaxation of up to ~35 Hz. At this temperature, the C-terminal D62 has a differential rate of almost 100 Hz.

Figure 4.25 displays the resulting picture of differential relaxation in case of a rigid and a mobile residue. TROSY spectra with selection of the narrowest and the broadest component of the quad images are shown for V53 and G5. While the difference in signal intensity and linewidth is marginal for V53, G5 almost disappears upon swapping of the selected coherences. The cross-section on the right side of each picture was generated in parallel to the ^{15}N axis in each case.

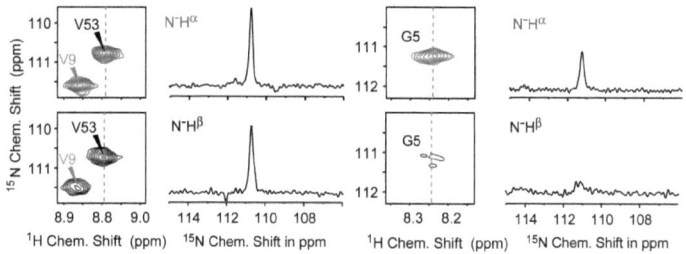

Figure 4.25. Comparison of narrow and broad HN multiplet components in a spin state selective H/N-correlation.[155] While inflexible residues (represented by V53, left) do not display pronounced differential relaxation, residues in flexible regions (e. g. G5, right) display a substantial difference between the narrow and the broad component. The narrow component (coherences N^-H^α/ H^-N^α) is depicted in red, the broad component (N^-H^β/ H^-N^β) is shown in black (below). 1D traces represent slices along the ^{15}N axis (dashed lines). The spectra were recorded and plotted identically and without use of apodization. The sample contained 25 % protons in exchangeable sites and was spun at 24 kHz with an effective temperature of 22 °C.

Differential relaxation plays an even bigger role for experiments in which transverse magnetization is needed in order for magnetization transfer. This applies e. g. to scalar transfers like INEPT. For ^{15}N/^{13}CO, ^1J-couplings amount to approximately 15 Hz. For ^{15}N/^{13}C$^\alpha$ transfers, the respective coupling is on the order of 11 and 7 Hz for intra- and interresidual transfers, respectively. Accordingly, transverse magnetization decays during the respective dephasing/rephasing periods of ~24 ms. Although deuteration allows for low relaxation losses due to heteronuclear dipolar interactions, high relaxation rates due to slow motion can deteriorate the performance of INEPT based experiments in case of flexible residues. Particularly these residues, however, require that scalar transfer be used, since dipolar transfers are hampered by residue motion (vide ultra). In these cases, selection of the slowly relaxing coherence may provide significant improvement of the experiment. If this is true or not depends very much on the time scale of the motion. For C-terminal residue D62, Figure 4.26 displays three different motional scenarios by choice of different experimental temperatures. The signal intensity after 48 ms ^{15}N transverse magnetization is compared to that of a rigid residue in the cases of a coherence selective and a standard magnetization transfer. The employed delay time is the time usually encountered in scalar triple-resonance out-and-back sequences.

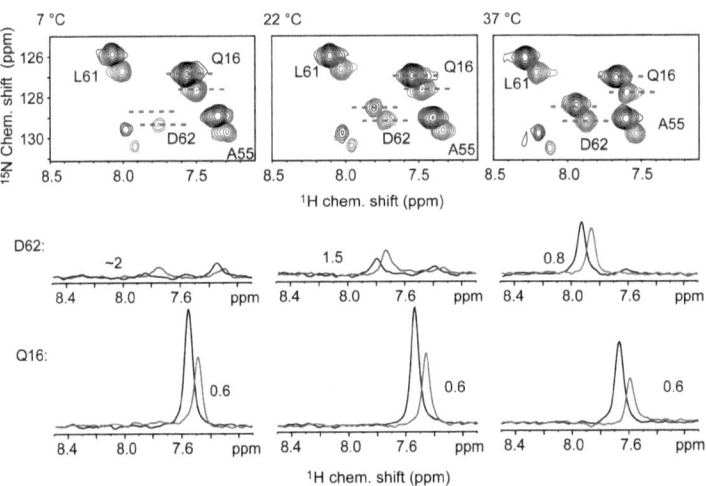

Figure 4.26. H/N correlation with a ^{15}N-T_2 filter. The filter-time was set to 48 ms, which corresponds to the typical scalar coupling evolution periods in triple-resonance HNCO type experiments. Correlations with mixing of the spin states are depicted in black, and those including coherence selection are shown in red. For quantitative comparison, 1D cross-sections along the proton dimension (as indicated in the 2D by dashed lines) are represented at the bottom of the Figure (for Q16, D62). Depending on the temperature (and thus the amplitude of internal motion), spin state selection results in better signal to noise for flexible residues like D62. The factors depicted in the Figures reflect the ratio between coherence selective and standard experiment. The indicated temperatures represents the effective temperatures. TROSY peaks are shifted by $J_{NH}/2$ in both dimensions. Spectra were obtained using a deuterated sample which was prepared using 10 % H$_2$O in the crystallization buffer. The MAS frequency was set to 24 kHz at a ^{1}H Larmor frequency of 400 MHz.

Methodological work on the SH3 domain

Triple-resonance experiments including an ST2-PT were introduced for solution already in the late 1990s.[156,157] The original pulse sequences for solution can be adapted to the solid state according to the description given in Chapter 4.2. As shown in Figure 4.27, severe differences in signal intensity are found for residues undergoing slow motion. Rigid residues loose in intensity when compared to the non-selective experiment. The double-resonance experiments (HSQC and TROSY, depicted in A and B for completeness, respectively) provide only little gain for mobile residues, while rigid ones loose approximately half of the intensity of a non-selective experiment. For the triple-resonance experiment, however, the difference between rigid and mobile residues is critically enlarged. An intensity gain of a factor of 1.5, 1.8, 1.9, and ~3 e. g. apply for G5, E45, R49, and E7, respectively, according to a factor 2.3, 3.2, 3.6, and 9 in measurement time. Comparable values account for the other (even weaker) residues annotated in bold in Figure 4.27, which, however, are compromized by lower overall signal to noise or peak overlap. All of these experiments were recorded at 700 MHz. Larger gains will be possible at higher fields.

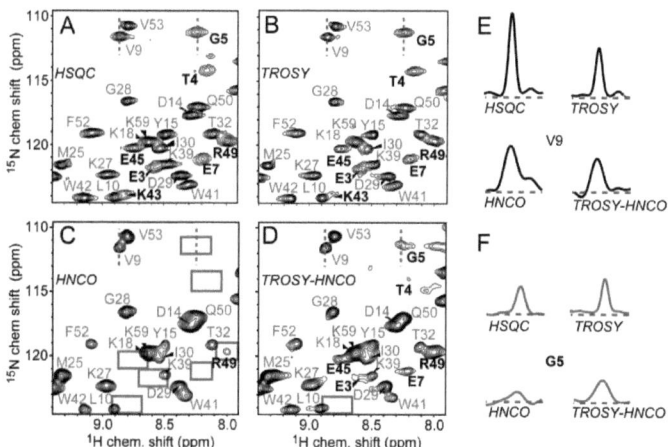

Figure 4.27. HN correlations recorded with an HSQC (**A**), TROSY (**B**), standard HNCO (**C**) and TROSY-HNCO scheme (**D**), recorded as a 2D without ^{13}CO incrementation. The intensity of mobile residues, which appear as broad resonances in the HSQC, decreases in the triple resonance experiments due to relaxation during the N-CO INEPT period. This problem is decreased by use of coherence selection. Respective residues are annotated in bold, grey labels indicate immobile residues. Blue boxes indicate resonances below the lowest contour. The double resonance experiments in A and B were recorded and processed under identical conditions and plotted to the same scale. The same is true for HNCO (C) and TROSY-HNCO (D). The

Methodological work on the SH3 domain

lowest contour was chosen as 3σ compared to noise RMS. (Triple-resonance experiments in C and D were recorded and processed using parameters different from A and B.) **E)** and **F)** show ^{15}N sections at resonance positions of V9 and G5, serving as representatives for normal and mobile residues, respectively.

The general loss of signal to noise in the absence of differential relaxation is due to the fact that only one half of the initial ^{1}H magnetization contributes to the selected coherence, while the other half of the magnetization is sacrificed. Additionally, the experiment using ST2-PT transfer is slightly longer and uses more pulses, which are intrinsically imperfect and thus prone to loss of signal to noise. On the other hand, TROSY is sensitivity-enhanced in terms of a preservation of equivalent pathways[158], which again improves the signal to noise by a factor of ~√2 in multidimensional experiments.

For a quantitative analysis of the intensity gain, Figure 4.28 gives a composition of the signal intensities obtained in the four spectra in Figure 4.27. Double- and triple-resonance experiments were recorded and processed using different conditions.

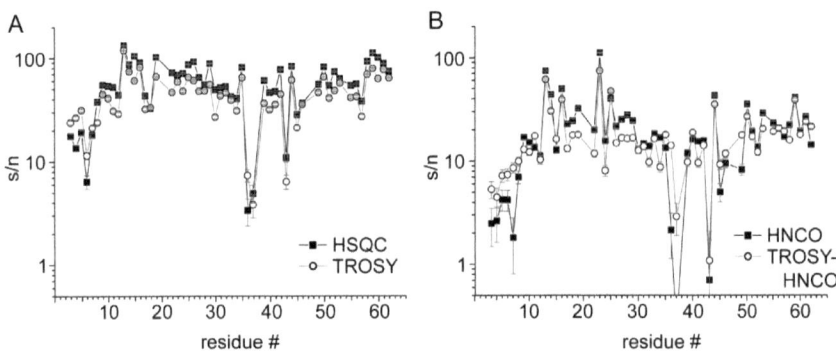

Figure 4.28. Comparison of relative intensities in an HSQC, TROSY, HNCO, and a TROSY-HNCO. Although the overall performance of the coherence selective experiments is worse, mobile residues are critically improved in the TROSY-experiments. This applies significantly more strongly to the triple-resonance experiment. **A)** Site-specific signal to noise of an HSQC (black squares) and a TROSY experiment (white circles). Both spectra were acquired recording 250 complex points according to 65 ms t_{1max}, using the same acquisition and processing parameters. For apodization of the direct dimension, 5 Hz exponential line-broadening was applied. **B)** Signal to noise of an HNCO (black squares) and a TROSY-HNCO (white circles), recorded as an H/N-2D without ^{13}C evolution. Both experiments were recorded and processed under

identical conditions with 70 complex points in t_1, corresponding to 22 ms t_{1max}. Procession was performed including 30 Hz exponential multiplication. All spectra were recorded within 2 h at 22 °C effective temperature and 24 kHz MAS at 600 MHz ^{1}H Larmor frequency. For the indirect dimensions, Gaussian Multiplication with a line-broadening of 10 Hz and a shift of the bell by 0.1 was used.

By the help of INEPT and ST2-PT transfers, flexible parts of the SH3 domain can be observed in multidimensional solid state experiments for the first time. The implications for a complete assignment of the protein backbone of this protein is described in Chapter 4.6. Generally speaking, the gain for residues undergoing slow motion upon use of coherence selection tends to be crucial especially for detection and assignment of protein parts that are important for protein functionality, like e. g. active centers of enzymes.

4.4 Side chain assignment

(Part of the work in this chapter was published in publication #12 on page 183)

Backbone assignment is an important first step in protein structure calculation. The protein residues' side chains, however, contain a number of atoms an order of magnitude larger than present in the backbone. Especially, ^{13}C resonances are important for distance restraints between different amino acids in the hydrophobic core. Furthermore, an assignment can be largely facilitated by the knowledge of the amino acid type, which can be determined by the help of side chain chemical shifts.

Full side chain correlations for deuterated proteins have been obtained for hydrophobic amino acids using residual protonation in deuterated proteins.[159] In order for a complete side chain assignment of deuterated proteins, correlations to the amide backbone would be necessary in order to include residues without methyl groups. Here, the strategy of an out-and-back transfer starting and ending with ^{1}H^N magnetization as in the case of the backbone assignment experiments is not viable anymore, since too many successive magnetization transfer steps would hamper a reasonable signal to noise. Instead, starting magnetization in the side chain would be favourable for a one-way transfer to the amide group. In principle, direct cross polarization from ^{1}H can be used to excite carbons. This, however, seems inconvenient for long side chains, where ^{1}H magnetization of only one amide will be split up for excitation of ^{13}C nuclei. A distribution of C$^\alpha$ magnetization along the side chain before ^{13}C chemical shift evolution results in poor polarization of aliphatic carbon nuclei. This can easily be probed by a comparison of the ^{13}C one pulse direct excitation spectrum with the 1D ^{13}C spectrum resulting from Cross Polarization from ^{1}H or ^{1}H/^{15}N with a subsequent mixing step.

Direct excitation of ^{13}C is usually inconvenient not only due to a four times lower gyromagnetic ratio in comparison to protons but also because of an extremely long recycle delay, which is even longer in the absence of protons in deuterated samples.[46] Due to a widely enhanced longitudinal relaxation by use of Paramagnetic Relaxation Enhancement (PRE) (see details in Chapter 3), experiments that are based on a direct excitation of ^{13}C nuclei can be performed under reasonable recycling. Figure 4.29 shows inversion recovery curves for different ^{13}C bulk resonances in the presence of 75 mM Cu(edta) in a protein sample with 25 % protonation in exchangeable sites. Fitting of the data yields T_1 values of 1.0 s for the aliphatic region, 1.6 s for the C^α-region, and 1.4 s for the CO bulk.

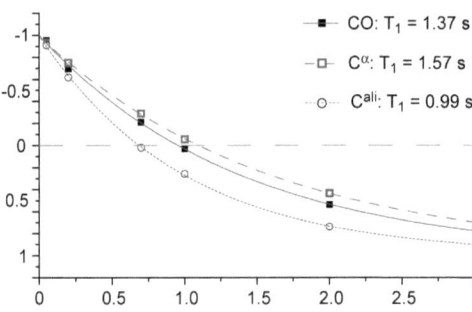

Figure 4.29. Inversion recovery data for CO, C^α, and aliphatic carbons under 75 mM Cu(edta)-doping in the presence of 25 % protons in exchangeable sites. Aliphatic carbons return to equilibrium with a T_1 of 1s, whereas CO and C^α have a T_1 of 1.37 and 1.57 s, respectively.

Full-side chain correlations

An experimental scheme for the acquisition of a full-side chain correlation is shown in Figure 4.30. The strategy of employing an adiabatic mixing sequence in the experiment (TOBSY-mixing[160,161]) goes back to work done by Agarwal et al..[62] Since the C/H transfer is partially bidirectional, the obtained spectra also yield sequential correlations and can in principle be used for a sequential assignment. This feature of long-range C/H transfers is used in the following section for a ^{1}H shift based sequential assignment. For experiments yielding intra-residual ^{13}C side chain correlations, a direct transfer of ^{13}C magnetization to ^{15}N can be used in principle. This, however, results in comparably lower signal to noise for non-C^α-resonances than with the detour pathway along ^{1}H, as probed by the ^{13}C detected reverse experimental scheme. This effect is caused by the non-specific C/H transfer, which provides an effective $C^{ali.}/H^N$ contact already directly in the absence of intermediate ^{13}C mixing. These desired long-range contacts can also be enforced for CN-CPs with longer duration. However, even for a CP duration of 14 ms, the obtained long range contacts were observed to be only half as strong as the contacts obtained with H/C long range transfers of 3 ms transfer time.

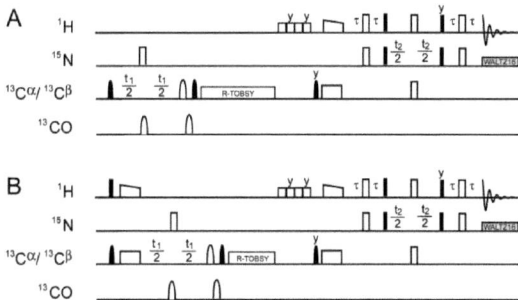

Figure 4.30. Pulse scheme for a full side chain CNH-correlation for sequential backbone assignment. By bi-directional transfer to $^1H^N$, both, inter- and intra-residual cross peaks can in principle be observed. A) Direct excitation of ^{13}C resonances. B) Use of direct excitation in combination with CP enhancement. Both experiments employ Tobsy-mixing[160,161] as proposed for deuterated samples by Agarval et al..[62]

For the start magnetization on ^{13}C, the natural polarization can be used in terms of a direct excitation. As an alternative, a CP polarization transfer from ^{1}H can be added to the natural ^{13}C polarization. This works since the CP pulse on the ^{13}C channel spin locks the coherence gained by natural ^{13}C polarization. The phases have to be chosen such that the relative phases of the first four pulses do not change in the course of the phase cycle and TPPI incrementation. The effect of a magnetization add-up is represented in Figure 4.31 (obtained by leaving out respective pulse scheme elements). As seen from the comparison between blue and green spectra, the signal from ^{13}C natural polarization is slightly reduced by the spin lock. The overall signal intensity, however, is increased by the increased starting magnetization (of mainly C^α and C^β).

Methodological work on the SH3 domain

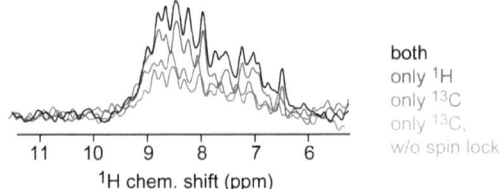

Figure 4.31. Comparison of the different contributions to the observed overall signal. The spectra were recorded without ^{13}C mixing, which would decrease the relative improvement of the CP sustainment by an equalisation of different start magnetization.

A representative part of the side chain correlations obtained with the pulse scheme A in Figure 4.30 on the SH3 domain of α-spectrin is shown in Figure 4.32. Interresidual contacts only appear in part of the strips, since the transfer efficiency is relatively small in comparison to the intraresidual one.

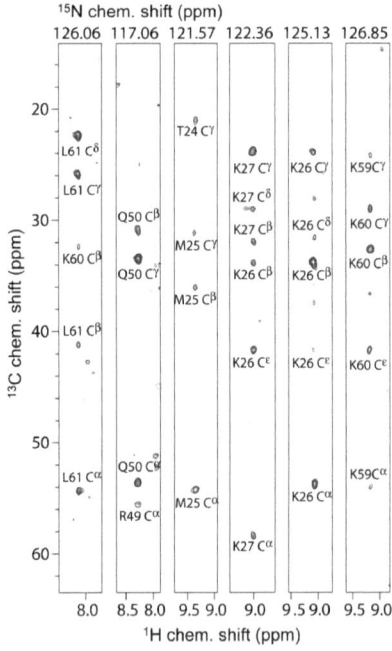

Figure 4.32. Exemplary subset of full-side chain correlations of the SH3 domain of α-spectrin, recorded with the pulse scheme in Figure 4.30 A, using 14.4 ms Tobsy-mixing.[160,161] In principle, by a bi-directional

Methodological work on the SH3 domain

C/H-transfer, the employed pulse scheme allows for a sequential assignment. This is, however, only true for a fraction of the resonances.

Full side chain correlations are valuable information not only for a determination of the amino acid type of the residue and the assignment of side chain chemical shifts. In addition to backbone and side chain assignment concerning ^{13}C resonances, assignment of the NH_2-groups is crucial for a complete understanding of protein function. Assignment of amide moieties with high reliability is important since amide groups of the Asn and the Gln side chain tend to be involved in salt bridges stabilizing the tertiary or quaternary structure of biomolecules. Figure 4.33 shows the unambiguous assignment of the side chain amide of Q16 by the help of full side chain correlation experiments. Here, the assignment can in principle be derived from an HNCACB spectrum. The signal intensity is sufficient for mapping the C^β resonance in the strip with the ^{1}H and ^{15}N backbone resonances to the strip with the ^{1}H and ^{15}N shifts of the side chain amide moiety. This single match, however, can be ambiguous in case of larger proteins. For additional clarity, the whole side chain experiment (shown in blue) can be mapped to the HNCACB strips. Besides the C^β resonance, the C^γ resonance, which matches the side chain strip, shows up in addition. The unambiguity achieved by this approach very much facilitates a complete assignment of side chain residues.

Methodological work on the SH3 domain

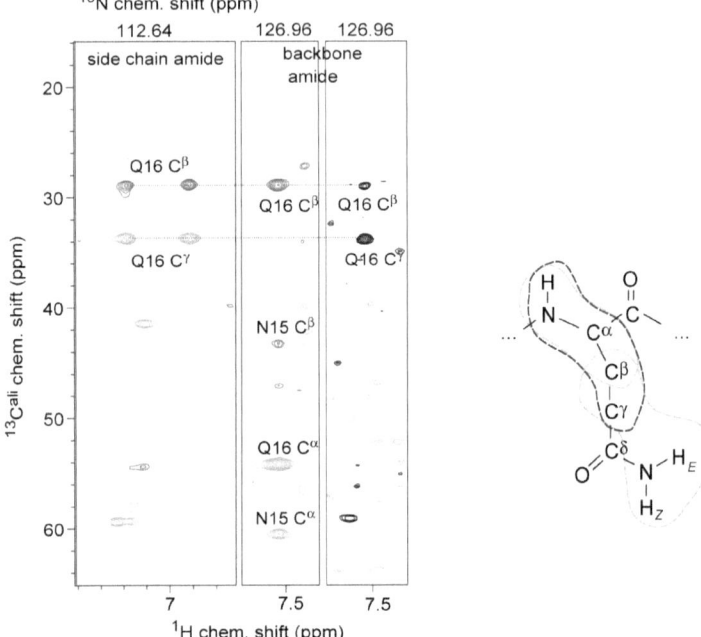

Figure 4.33. Exemplary assignment of a side chain amide. The mere match of C^β in the HNCACB strips leads to ambiguities when more Gln/Asp side chain amides are to be assigned. Unambiguous assignment by a pair of matching frequencies (C^β and C^γ) can lead to higher reliability. In this case the Q16 C^α resonance does not show up in the full side chain experiment. For backbone amides frequencies, the HNCACB spectra (see above) displays C^α and C^β resonances in light and dark green. At the right side, shifts correlated by the full side chain correlation (blue) and HNCACB (green) are illustrated.

Side chain amide correlations

Besides an electrostatic stabilization important for tertiary structure formation, side chain amide groups are responsible for a solvation of the protein or protein-ligand contacts in the course of signal transduction or target recognition. In principle, information about the NH_2-groups can be obtained by mapping ^{13}C chemical shifts correlated to the amide H/N shifts to those that have been assigned in the backbone sequential walk. Due to high solvation for most side chain amides, however, proton exchange of the amide group with the surrounding solvent is an issue for the respective resonances. Especially the INEPT-based out-and-back experiments rely on a high kinetic stability of the involved proton, since they have an overall duration on the order of 70 ms. In addition, the success of these experiments is dependent on the motional characteristics of the involved residues. While for residues pointing out to the solvent, a high flexibility is intuitive, those involved in salt bridges are prone to undergo slow motional processes, which complicates acquisition by unfavourable fast relaxation (vide ultra).

Based on the idea of a ^{13}C direct excitation according to the description above, an experiment for the evolution of $^{13}C_{ali,\,sc}$ and $^{13}CO_{sc}$ chemical shift in the course of two indirect dimensions in addition to a direct ^{1}H acquisition was created. The pulse scheme of this CACOH is presented in Figure 4.34. The INEPT-based transfer among the involved ^{13}C nuclei was chosen instead of a dipolar mixing in order for a clean and directed transfer among the carbons. Polarization transfer to ^{1}H can be achieved by use of a long-range H/C CP. Water suppression is affected according to Zhou et al..[162]

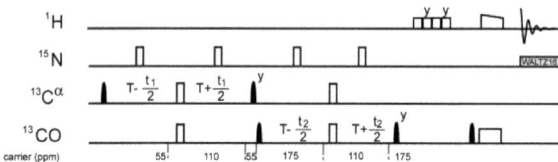

Figure 4.34. Pulse scheme for a 3-dimensional ^{13}C direct excitation based C/H correlation experiment. Direct excitation of $^{13}C^{ali}$ is followed by a refocused INEPT to $^{13}CO_{sc}$. A long-range transfer to ^{1}H is achieved by a 1.5 ms CP (70 and 45 kHz field strength for the ^{1}H and ^{13}C channel). Water suppression was achieved as denoted above. Open and closed bars refer to 180 and 90° pulses, respectively. Selective pulses were applied using on-resonance soft rectangular pulses. The according ^{13}C frequency (in ppm) is denoted in the bottom of the Figure. T is set to 3 ms.

For the homonuclear ^{13}C transfer, the delay time T in the refocused INEPT takes into account the evolution of competing ^1J-couplings to the adjacent aliphatic (as a passive coupling) and to the carbonyl carbon (as the active coupling) with a coupling constant ^1J of 35 and 55 Hz, respectively. The relative signal intensity of the transfer for the first INEPT can be calculated as

$$I(T) = \sin(\pi\,^1J_{ali\text{-}CO}\,T)\cos(\pi\,^1J_{ali\text{-}ali}\,T)\exp(-2T/T_2). \tag{74}$$

Figure 4.35 shows a simulation of the transfer efficiency dependent of the duration of the INEPT for a T_2 of 40 ms. The ideal transfer amplitude can be reached with an INEPT duration of 2T = 6 ms. Dashed lines represent the evolution of the passive and active coupling alone.

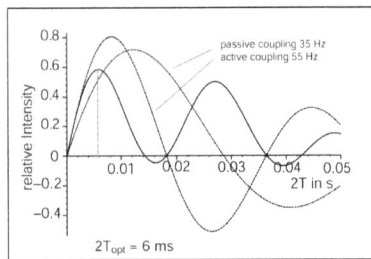

Figure 4.35. Transfer efficiency for the $^{13}C_{ali, sc}/^{13}CO_{sc}$ refocused INEPT in the CACOH experiment (solid black line). Due to simultaneous evolution of the passive coupling to the vicinal aliphatic carbon, the transfer efficiency is compromised and reaches a maximal value at only 6 ms duration. For the simulation, a T_2 relaxation time of 40 ms was assumed.

Using this pulse scheme for the observation of side chain amides, additional correlation signals show up in comparison to the described out-and-back experiments. Figure 4.36 A shows an overlay of the $^1H/^{13}C$ projection with the respective projection of a 3D out-and-back HNCA. Grey boxes indicate the side chain amide signals. An according comparison of the $^1H/^{13}CO$ projection to HNCO and HNCACO projections is given Figure 4.36 B and C. For all out-and-back experiment, the signals arising from side chain amides are weak in comparison to the direct excitation CACOH experiment. The appearance of CACOH signals overlaying with either of both out-and-back correlations dues to a non-selective transfer of CO_i magnetization to inter- and intra-residual amide protons ^1H_i and $^1H_{i+1}$. In some cases, signals present in the out-and-back experiments do not show up in the direct excitation approach. This is probably due to the use of a dipolar magnetization transfer in case of mobile residues or a short T_2 ($^{13}C^\alpha$).

Methodological work on the SH3 domain

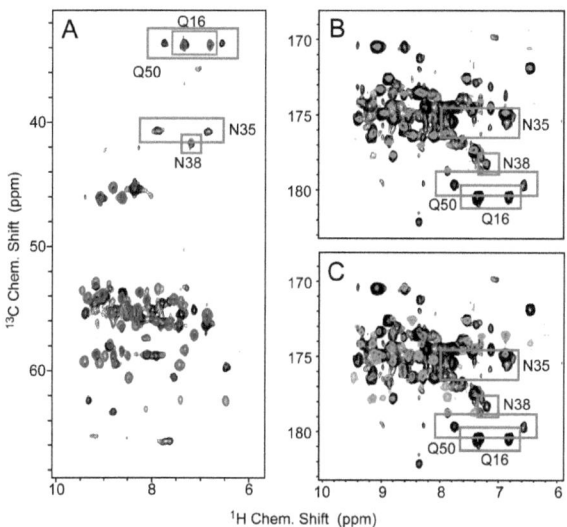

Figure 4.36. Comparison of side chain signals in the CACOH experiment to triple-resonance out-and-back experiments. **A)** Overlay of the $^1H/^{13}C^{ali}$ projection with an H/C^α projection of an HNCA (red). **B)** and **C)** Overlay of the $^1H/^{13}CO^{sc}$ projection of the CACOH with the respective projection of an HNCO (green) and an HNCACO (blue), respectively. The experiments were recorded and processed in a comparable manner.

The implications for the assignment of amide side chain amide moieties of the SH3 domain of α-spectrin are described in Chapter 4.6. In order to give an exemplary representation of the assignment process, Figure 4.37 shows the combination of respective strips of a CACOH and the HNCACB for the side chain and the HNCACB strip for the backbone resonances in case of N35. Here, both side chain carbons, C^α and C^β, show up in the HNCACB strip through one of the amide resonances and can be mapped to the strip through the respective backbone amide. However, the signal is hardly higher than the noise level and the resonances with the chemical shifts of the other amide are not visible. The CACOH experiment, on the other hand, contains both correlations with a good signal to noise. Due to the direct transfer of magnetization from ^{13}C to the detection nucleus, fast signal decay on the ^{15}N is prevented and ^1H exchange in the course of the transfer and evolution periods does not interfere with the signal to noise of the experiment. In this case, assignment of the (E)-proton of the amide moiety could only be assigned by use of the direct excitation experiment.

Methodological work on the SH3 domain

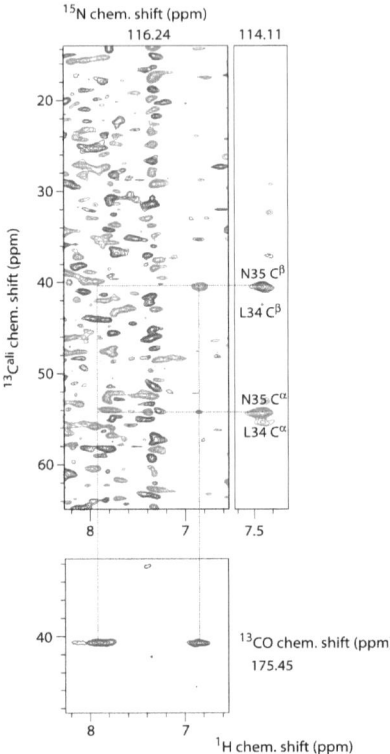

Figure 4.37. Support of the side chain amide signal assignment by the ^{13}C direct excitation experiment. Due to a ^{13}C starting magnetization, no exchange of the side chain amide protons can occur in the course of the transfer. Accordingly, also proton correlations are seen that have pronounced exchange with the bulk solvent or decay quickly due to large T_2 (^{15}N) due to motion.

In addition to the described side chain assignment, the CACOH experiment gives a complementary approach for a backbone assignment of dilute protonated proteins. By choosing the C/H Cross Polarization step such that inter- and intra-residual transfers are of the same efficiency, a sequential assignment of the backbone can be achieved. The transfer amplitudes are on the same order, since the distances between ^{13}CO$_i$ and ^{1}H$_i$ (2.6 Å) and ^{1}H$_{i+1}$ (2.0 Å) are comparable. Figure 4.38 shows a representative part of the sequential backbone walk, including residues 22-32. Although signal to noise is sufficient, the resolution is determined by the length of the constant-time delay within the

C/C INEPT. For this reason, many signals show up "bleeding through" from planes with different ^{13}CO shift in the strips in Figure 4.38.

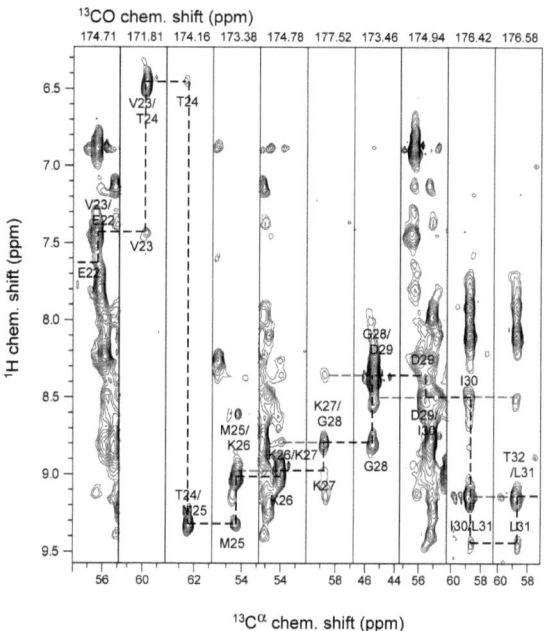

Figure 4.38. Walk on the backbone by sequential CO/H contacts. For each C^α-CO pair, magnetization is transferred to an inter- and an intra-residual ^1H. Due to the shorter distance to $^1H_{i+1}$, the inter-residual contact is stronger in each of the strips. Due to low resolution in the ^{13}CO dimension, signals from other planes bleed through. In addition to the strong sequential correlations (thick black lines), non-sequential correlations (depicted with thinner lines) to other protons close in space can be found. This is true e.g. for the kink in between K26 and I30 (vide infra).

Since the $^{13}CO/^1H$ magnetization transfer occurs through space, contacts to protons across the β-sheet can be found in addition to sequential backbone contacts. This is particularly useful, since these contacts are important for the determination of the overall fold of the protein. Figure 4.39 A displays several long-range contacts between CO and ^1H nuclei. D29-^1H is in contact to A11-CO (4.66 Å), I30-^1H makes a correlation to G28-CO (4.24 Å). K26-CO has a correlation to G28-^1H (4.07 Å) and to D29-^1H (3.10 Å). In this particular kink, relatively many non-sequential contacts

Methodological work on the SH3 domain

are seen due to the bend structure of the sequential amino acids. The residue pair V44/G51 is a representative for a usual β-sheet. Here, both possible CO/^{1}H-contacts are observed (V44-CO/G51-^{1}H: 3.22 Å, V44-^{1}H/G51-CO: 3.01 Å). Additionally, G51-CO gives a correlation signal with V53-^{1}H (4.20 Å). The respective distances for the case of the kink around D29 are depicted in Figure 4.39 B. It is expected that many more long-range contacts are present. Mostly, however, resonances are not sufficiently resolved in the third dimension.

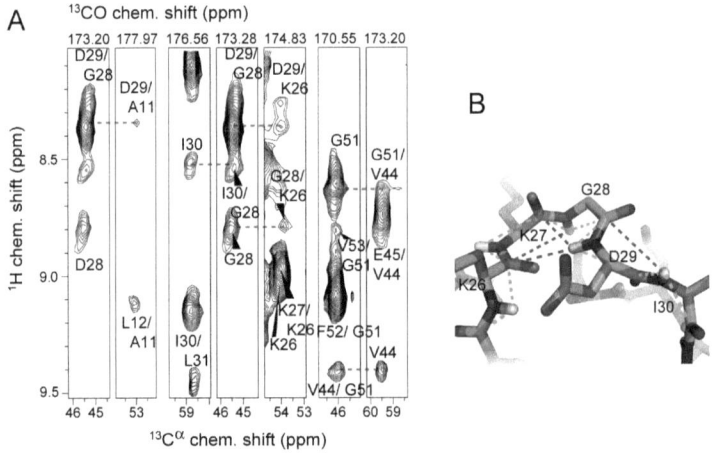

Figure 4.39. Non-sequential contacts arising from the CO-H magnetization transfer through space (**A**). In the kink around G28, which is structurally represented in (**B**), many long-range transfers can be seen. Blue dashed lines denote these non-sequential distances, orange ones depict the intra- and inter-residual sequential contacts. The last two strips in (**A**) represent the mutual contact of residues V44 and G51.

4.5 Magnetization transfer through space – proton RFDR experiments

(Work in this chapter was published in publication #11 on page 183)

In order to get to a structure of a protein, the mere chemical shift information, which can be extracted after sequential assignment, already gives hints about the secondary structure of protein domains.[163] Primarily, $^{13}C^\alpha$ and $^{13}C^\beta$ chemical shifts can be translated into so called secondary chemical shifts.[164] These are characteristic for α-helices (with C^α chemical shifts higher and C^β chemical shifts lower than random coil structure) and β-sheets (with C^α chemical shifts lower and

C^{β} chemical shifts higher than random coil structure). The respective information are usually taken into account in structure calculations by the help of the program TALOS.[165] Nevertheless, direct distance restraints between nuclei are important for a correct local fold. While in solution NMR, the Nuclear Overhauser Effect (NOE) is used for magnetization transfer through space, solid state NMR can employ dipolar interaction to obtain through-space transfer. In protonated solids, even fast MAS (up to ~60 kHz) does not achieve a full averaging of ^{1}H/^{1}H dipolar couplings. Many sequences aiming for an effective exchange of magnetization focus on this interaction. Examples for a mixing using proton dipolar couplings are DARR[49], DREAM[50], or PDSD.[166] Often, distance restraints between heteronuclei can be obtained by indirect use of the proton bath via a back and forth-transfer to protons for an efficient mixing.[167] Direct recoupling of heteronuclear dipolar interaction can be performed by Radio-Frequency Driven Recoupling (RFDR), in which a non-zero dipolar average Hamiltonian is created by rotor-synchronized 180°-pulses on the respective nucleus.[43,44] For a dilute proton content, an active recoupling of ^{1}H/^{1}H dipolar interaction as in the case of heteronuclei can yield distance restraints. This has been shown for ubiquitin, which structure could be refined by inclusion of direct proton-proton distance restraints to a backbone RMSD of 0.82±0.14 Å.[58]

An assignment approach for a mutual correlation of amide moieties has been proposed earlier.[168,169] Instead of a stepwise magnetization transfer through the backbone, we can use the through-space transfer in order to obtain chemical shifts of one amide moiety in addition to the chemical shifts of the adjacent one. Given the fact, that sequential amide moieties are in close spatial proximity, a dipolar HN-HN-correlation should also enable a sequential assignment without involvement of ^{13}C nuclei. Such, the HN/HN-correlation can be helpful for distance restraints as well as for ambiguities in the assignment.

Experiments providing correlations exclusively between amide moieties were established already in early biomolecular work with protonated systems in solution.[170,171] These NOE-based sequences can be applied to the solid state if only the mixing step of the pulse sequence is adapted to the above mentioned mechanism. The combination of a ^{15}N edited and a ^{1}H edited correlation experiment gives the same information as a respective 4D experiment in which both dimensions are evolved. The information can be extracted more accurately from a 4D experiment, this, however, would result in a $2^{0.5}$ fold loss of an additional dimension and in the difficulty of reaching sufficient resolution in a given time if no Reduced-Dimensionality (RD) approach[172] is employed. A solid state pulse scheme of a HSQC-NOESY-HSQC using RDFR as the element for proton-proton magnetization transfer is depicted in Figure 4.40 A. The magnetization transfer steps are achieved by ramped H/N CPs in each case. The first indirect dimension resolves ^{15}N chemical shift of the spatially close amide moiety by a simple back-and-forth transfer of ^{1}H starting magnetization. The second HSQC-type block is enriched by water suppression through four successive water purge pulses according to Zhou et al..[162] Heteronuclear decoupling from ^{1}H is achieved by use of Walz-16[132] with a power of 10 kHz. The RFDR block consists of rotor-synchronized π-pulses (one pulse per rotor period) on protons, using a pulse strength of 65 kHz. The mixing time was chosen to be between 3 and 15 ms (vide infra), corresponding to 24 to 360 cycles at 24 kHz MAS.

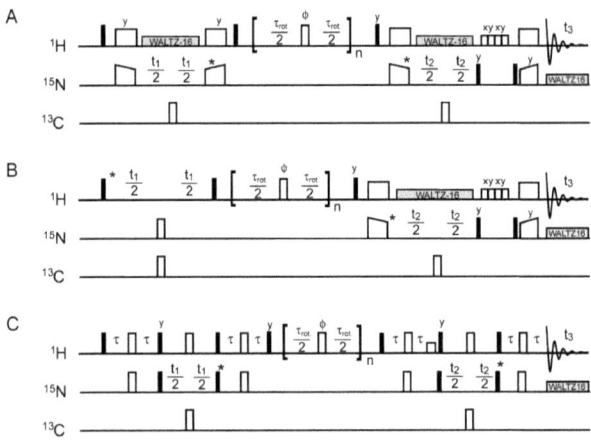

Figure 4.40. Pulse schemes for the correlations resolving spatial HN/HN contacts. (A) Pulse scheme for the N-NH experiment. (B) Pulse scheme for the H-NH experiment. (C) INEPT-based N-NH experiment. CPs were applied as rectangular pulses on ^{1}H and 75-100% ramps on ^{15}N. Open and closed bars denote 180 and 90° pulses. The single open pulse of reduced height is a water purge pulse of 1 ms duration applied at a strength of 60 kHz. The block of 4 successive smaller open bars are applied at 15 kHz and a duration of 20 ms each. Pulses marked with an asterisk denote phase-sensitive incrementation according to TPPI. The phase of the 180° pulses (φ) in the RFDR-block is incremented according to XY-8. τ$_{rot}$ is the rotor period. τ is a delay of $^1J_{HN}$/4.

Methodological work on the SH3 domain

The counterpart for the pseudo-4D approach, the NOESY-HSQC, resolves the respective ^{1}H chemical shift before the mixing sequence. This experiment is suitable for contacts to methyl protons, for which reason the spectral width has to be chosen accordingly and a sufficient resolution requires relatively many points in the ^{1}H dimension. This pulse scheme is depicted in Figure 4.40 B. Figure 4.40 C gives the according HSQC-NOESY-HSQC from solution state, which is based on INEPT-transfers instead of Cross Polarization, adapted to solid state proton detection by the RFDR mixing sequence.

Cross peak build-up

Figure 4.41 shows the build-up of cross-peak intensity for an exemplary residue (Y13) for mixing times of 3, 8, and 15 ms. The residue was chosen such that little overlap in the additional ^{15}N dimension (t_1) appeared even for the longest mixing duration.

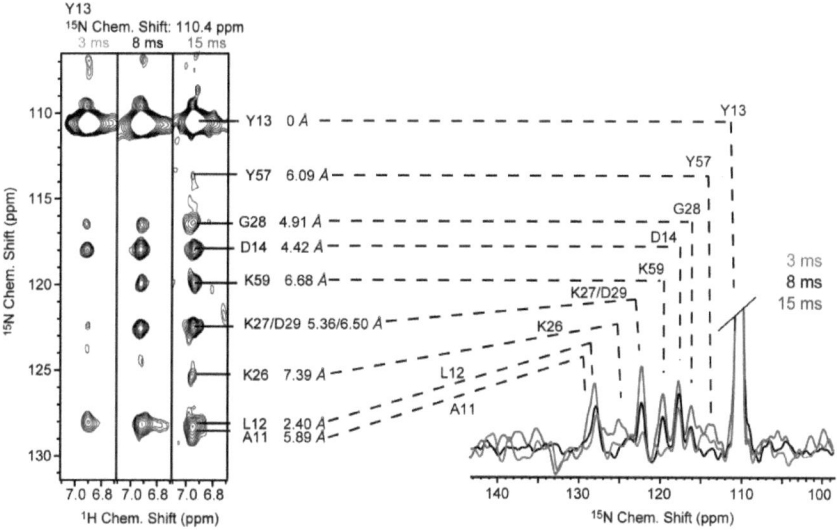

Figure 4.41. Build-up of magnetization upon RFDR mixing on protons. Y13 is taken as an example of the magnetization equilibration among vicinal amide sites. Successive spectra were recorded with mixing times of 3, 8, and 15 ms. Spectra were scaled to the same diagonal peak intensity. All spectra were recorded at 22 °C effective temperature and 24 kHz MAS using a deuterated sample with 25 % proton back-substitution in labile sites within 4 h each. t_{1max} amounted to 18 and 9 ms in the first and second indirect ^{15}N dimension. Distances are indicated in addition to the respective contacts of Y13 H^N.

Methodological work on the SH3 domain

At a mixing time of 3 ms, the sequential contacts (L12 and D14 with a distance of 2.4 Å and 4.4 Å, respectively) give the strongest cross peaks. The respective cross-peak intensity, however, is surprisingly comparable already at this mixing time despite the distance difference. Thus, in order to resolve short contacts in terms of a build-up, even shorter mixing times seem to be necessary. Also G28 (4.9 Å) and K27 (5.4 Å) have some discernable cross peak intensity even at a mixing time of 3 ms. Their reduced intensity accords to the increasing distances. At 8 ms mixing time, the proximal contacts give rise to cross peaks of an unchanged intensity, however, especially K27 with its longer distance results in a cross-peak twice as strong as at 3 ms. An additional contribution of D29 (6.5 Å) is responsible for this relatively strong intensity gain. Furthermore, K59 (6.7 Å) shows an intensity significantly larger than the noise level at 8 ms. For 15 ms RFDR, even long-range contacts like K26 (7.4 Å) give rise to very weak signals (with a signal to noise of 1.5). The general cross peak intensity in comparison to the diagonal peak is about 1/8. This value corresponds to 8 ms mixing and an intermediate distance (4 Å). The relative loss of intensity is in particular due to a probability of only ¼ for the adjacent amide moiety to bear a ^{1}H.

In order for a visualisation of the contacts that can be determined for Y13, Figure 4.42 displays the molecular surrounding as taken from the crystal structure at room temperature[60], with amide protons added with pymol.[74]

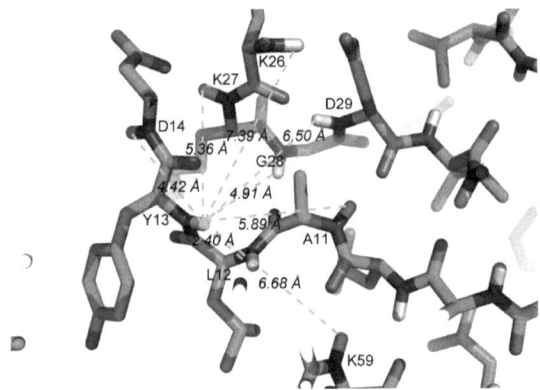

Figure 4.42. Molecular surrounding of Y13 with the adjacent amide moieties and their respective distances (annotated in italics) as a visualisation of the exemplary representation of cross peak intensity build up upon RFDR mixing. Y13 sits in a non-β-sheet region in the beginning of the RT-loop. The plot (including dis-

Methodological work on the SH3 domain

tance determination) was engineered with the help of pymol[74] from the X-ray structure published in Chevelkov et al..[60]

Figure 4.43 displays the build-up for additional two residues, namely G28 and L61. Since it is not clear how to normalize the build-up, the spectra with 15 ms mixing are shown with twofold normalization.

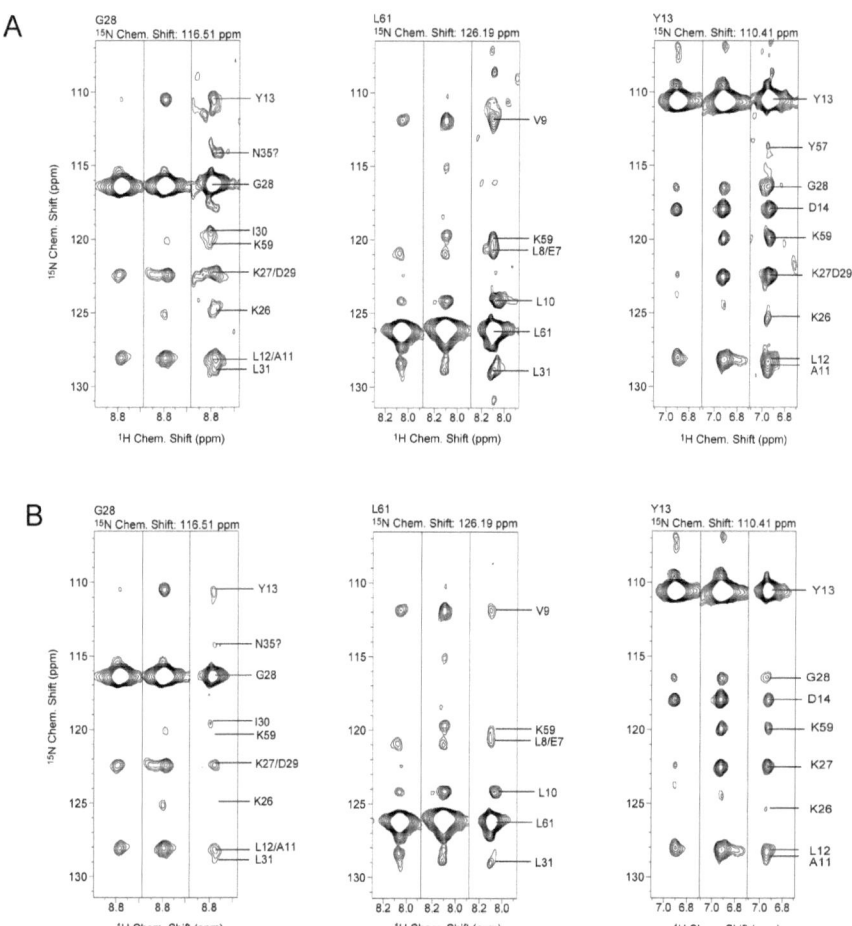

Figure 4.43. Build-up of magnetization upon RFDR mixing on protons. Besides Y13 (see Figure 4.41), two additional resonances are shown. For a quantification of the build-up, the question of how to calibrate the

Methodological work on the SH3 domain

intensity has to be answered. For this purpose, spectra were normalized according to their diagonal peak intensity (**A**) or to their signal to noise (**B**). In each case, the three strips resemble a mixing time of 3, 8, and 15 ms.

In order for an unambiguous extraction of neighbour-neighbour contacts and their cross-peak intensities, the combination of the N-NH and the H-NH experiment (corresponding to HSQC-NOESY-HSQC and NOESY-HSQC) can be used to generate the resolution of an alternative 4D spectrum. Figure 4.44 shows both strips taken for Y13 ^{1}H and ^{15}N chemical shifts in t_2 and t_3 plotted against the H/N-correlation recorded under identical conditions, but with a better resolution (70 ms t_{1max}). All contacts visible at a mixing time of 8 ms are indicated. The spectra were recorded with a t_{2max} of 9 ms and a t_{1max} of 18 ms and 20 ms in a total duration of 4 and 8 h in the case of the ^{15}N and the ^{1}H resolved experiment, respectively.

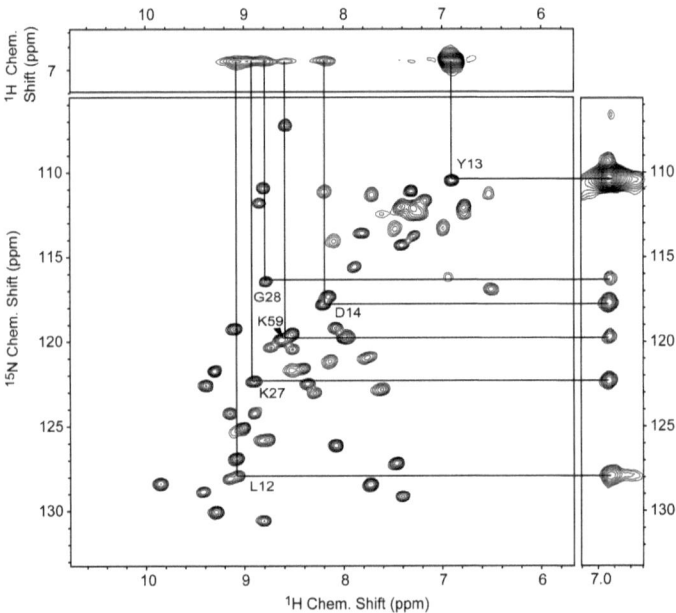

Figure 4.44. Combination of ^{15}N and ^{1}H resolved experiments (N-NH and H-NH) in order for an unambiguous assignment of spatial contacts. The spectra were recorded with 8 ms RFDR mixing. The respective t_{1max} amounted to 20 and 18 ms in the case of ^{1}H and ^{15}N resolved experiments. The spectra were recorded in 8 and 4 h, respectively. This reflects the need for sufficient resolution rather than the sensitivity (vide infra).

Methodological work on the SH3 domain

Backbone assignment by sequential contacts

Backbone assignment can be achieved by finding strips of sequential amino acids, which, in this case, is possible due to the relatively strong cross-peak intensities of the sequential contacts. The procedure can be done for the ^{1}H as well as for the ^{15}N resolved correlation experiment, such that ambiguities due to degeneracy in only one dimension are resolves as long as no overlap of signal occurs in the H/N-correlation. This results in an accuracy comparable to that of the HNCACB approach denoted above. Additional information results from the fact that sequential cross peaks of residue *i* exist for both directions, residues *i*+1 and *i*-1. A minor drawback, however, is the existence of the relatively strong diagonal signal, which in case of short evolution in the indirect dimensions additionally leads to artefacts due to truncation. Figure 4.45 represents a part of the sequential backbone walk for the SH3 domain of α-spectrin.

The ^{15}N resolved experiment was recorded as described above within 4 h, using 8 ms RFDR mixing. The ^{1}H based spectrum suffers from insufficient resolution, although signal to noise is adequate for the experiment. For these cases, in which resolution is the time limiting factor, non-linear sampling can be taken into account. The spectrum represented in Figure 4.45 was recorded with a minimal phase cycle of 2 and 200 and 50 complex points (8.3 and 9 ms) in the indirect dimensions t_1 and t_2 within 4 h. For procession, 200 additional points were linear predicted.

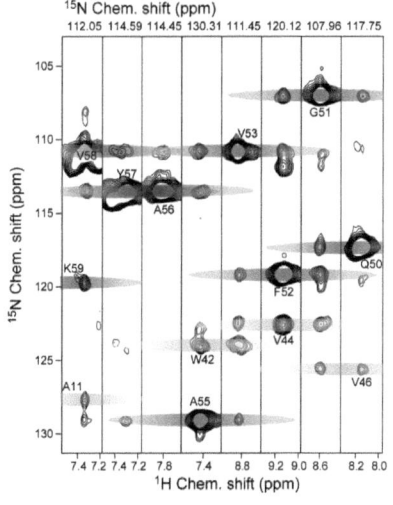
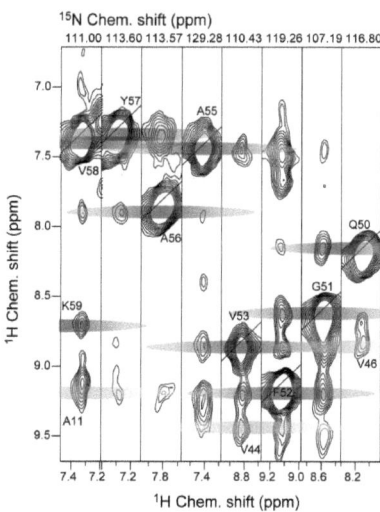

Methodological work on the SH3 domain

Figure 4.45. Sequential backbone walk for the SH3 domain of α-spectrin. Only a representative part (residues Q50 to V58) is shown. Besides the sequential contacts (marked in blue), information for the assignment is also contained in the spatial contacts to neighbouring β-sheets (marked in red). Both spectra were recorded in 4 h each.

Sequential peaks to residue *i* can be found choosing the respective chemical shift in F1 in both spectra and comparing the resulting F2/F3 H/N planes for identical cross peaks.

Solid state NOESY using scalar correlations
The assignment approach as well as the extraction of distance restraints are thought to be generally applicable. Most target proteins in solid state NMR do not crystallize as nicely as the SH3 domain in our case. For that reason, higher protonation levels and accordingly use of dipolar transfers will be preferred in comparison to scalar transfers. For flexible residues, the use of INEPT instead of CP, however, is crucial. Elevated amplitudes of motion can result in very low transfer efficiencies of CP steps (see Chapter 4.3).

The N-NH and H-NH experiments can therefore be modified such that H/N-transfers are affected like in solution state (see Figure 4.40 C for the respective pulse program). For INEPT transfer, a maximum efficiency of 100 % is obtainable if relaxation is negligible in the course of the transverse ^{1}H magnetization period (~ 5.5 ms). In the case of the sample with a 25 % proton content in exchangeable sites, the overall performance is a factor 1.5x better than with the use of Cross Polarization. This value refers to the bulk amide signal without RFDR mixing. However, water suppression is more difficult in comparison to the dipolar sequences due to predominance of H/N antiphase operators throughout the sequence. Figure 4.46 represents the bulk signal of the first slice for an HSQC, the H-NH-experiment, the N-NH-experiment and the N-NH-experiment using INEPTS using 128 scans in each case.

Methodological work on the SH3 domain

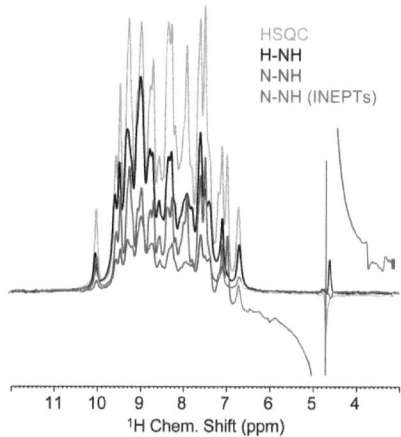

Figure 4.46. Comparison of the intensities obtainable with the experiments presented above in comparison to an HSQC. All spectra were recorded using 128 scans and a mixing period of 0 ms for a sample with a 25 % proton content in exchangeable sites. Relative intensities in comparison to the HSQC amount to 50 % (H-NH, black), 23 % (N-NH, red) and 37 % (N-NH with INEPTs, blue). Solvent suppression was achieved according to Zhou et al.[162] for H-NH and N-NH or with a 1 ms proton purge pulse for the INEPT-based experiment, respectively.

The main advantage of scalar based spatial correlations is the inclusion of mobile protein parts. Interestingly, cross peaks can be observed using dipolar based RFDR mixing although CP does not give any transfer for the respective residues. This is probably due to a recoupling of dipolar interactions by hard pulses, which does not require a constant coupling as in the case of Hartmann-Hahn-matching. As known from relaxation measurements (see Chapter 4.3), the N-terminus supposedly undergoes large amplitude motion. Consequently, these residues do not show up in the CP-based versions of the RFDR experiments. In contrast, they can successfully be correlated to their sequential neighbours using the INEPT version of the experiment. This is of interest, since these residues have not been assigned with traditional solid state NMR methods (see Chapter 4.6). Figure 4.47 gives a representation of the CP-invisible part of the RFDR based sequential walk. The spectrum was recorded and processed in analogy to the CP-version above.

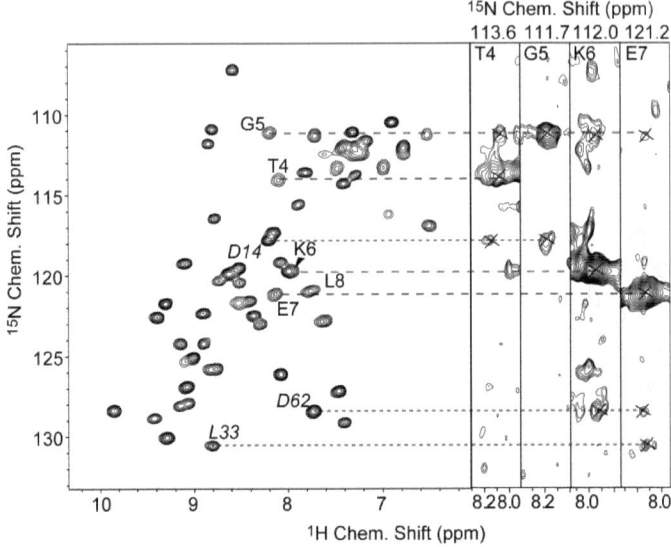

Figure 4.47. Assignment of the N-terminal part of the SH3 domain of chicken α-spectrin. The assignment of flexible residues can be performed by use of the INEPT-based versions of the RFDR experiments. The 3D spectrum at the right was recorded employing the pulse scheme in Figure 4.40 C within 4 h, using 18 and 9 ms indirect acquisition in t_1 and t_2. Negative contours due a low signal to noise are drawn using a single contour. The left side of the Figure represents an H/N correlation (see Figure 4.44), recorded in 10 min under identical conditions. Residues annotated in italics and correlated with smaller dashes are long-range contacts.

4.6 Application of improved methodology in ¹H detection to the SH3 domain of chicken α–spectrin

Although the SH3-domain of α-spectrin is used as a well characterized standard system for solid state NMR methodology in this work, a complete NMR characterization of this protein has not been achieved using the traditional ¹³C detected methods.[71,173] New insights for a complete NMR spectroscopic description of the SH3 domain of α-spectrin described in this Chapter are based on the methodological work described in the Chapters before. Although the focus of this work are the achievements of the methodology per se, the obtained information for this protein has a potential impact on further methodological studies based on the SH3-domain in the future.

Methodological work on the SH3 domain

The use of uniformly protonated samples has the advantage of a high abundance of different protons, serving as a carrier of starting magnetization on the on hand and giving structural information about the molecule on the other. However, even at fast MAS, the resolution obtainable with samples of this kind is at least a factor of 10 less than achievable when extensive deuteration is employed.[55,59] Figure 4.48 displays an H/N-correlation of the SH3 domain obtained with homonuclear frequency-switched Lee-Goldberg (FSLG) decoupling during indirect acquisition of protons, taken from van Rossum et al.[173]

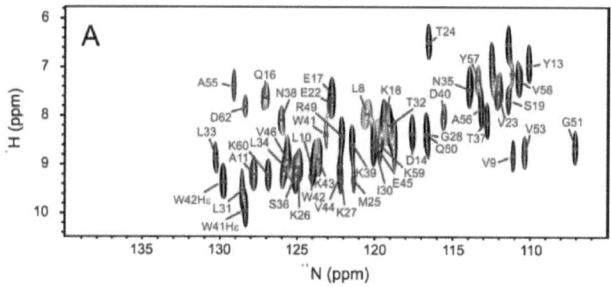

Figure 4.48. Resolution of the H/N correlation of the SH3 domain of α-spectrin obtainable with protonated protein samples, taken from van Rossum et al.[173]

Due to an predominant use of dipolar mixing transfer steps, mainly rigid parts of the β-sheet structure have been in the focus of previous investigation. A complete assignment of the protein, however, would be appreciable for the manifold dynamic[60,128,174] and methodological studies[59,62,72,129,141] that the SH3 domain is used for. Particularly the mobile regions at the N-terminus and the loop regions are interesting from this point of view. (see Figure 4.24 in Chapter 4.3. In addition, side chain resonances can be focused for a characterization of dynamics of different kinds. While fast rotameric jumps were observed for (side chain) methyl groups,[175] side chain amide groups are particularly rich in information about slow motional processes like chemical exchange.[174] The proton and ^{15}N chemical shifts of side chain amide groups could not be assigned in previous studies.[173,176] In addition, some $^{1}H/^{15}N$ chemical shifts could not be confirmed with triple-resonance out-and-back experiments. For other resonances, the assignment in the H/N-correlation was not clear due to resolution limits of the protonated samples. Figure 4.49 displays the compilation of unassigned or misassigned $^{1}H/^{15}N$ shifts.

Methodological work on the SH3 domain

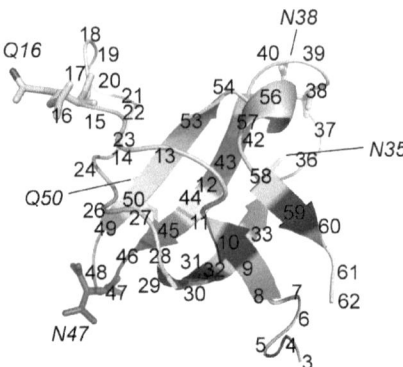

Figure 4.49. Residues with uncertain or incorrect assignments in previous studies (depicted in red). The N-terminus and the distal loop could not be assigned with ^{13}C detection based methods.[177] For the other red residues either a different chemical shift in either the proton or ^{15}N dimension was determined or the resolution was insufficient for clear separation of chemical shifts. The side chain amides of Gln and Asn residues (annotated in italics) were not assigned.

Reassigned resonances

Due to a manifold better resolution in the ^{1}H dimension, several resonances seemed degenerate in former approaches, but are significantly separated using extensive proton dilution. Such resonances can be assigned with the HNCA and HNCACB spectra or alternatively with the HNCACO experiment. This applies e. g. to resonances W42, L10, and K43, for which the ^{15}N chemical shift is almost identical. Another case with insufficient resolution in former studies is residue I30. Here, two non-confirmed assignments of proton chemical shifts for E45 and R49 make a transfer of previous assignments to our studies difficult. In our hands, R49 displays chemical shifts that deviate completely from former assignments. A reassignment of these three residues according to the triple resonance solution state experiments is shown in Figure 4.50. A match of C^{α} and C^{β} resonances with sequential residues V44, D29 and I30, and to Q50, respectively, can be found.

Methodological work on the SH3 domain

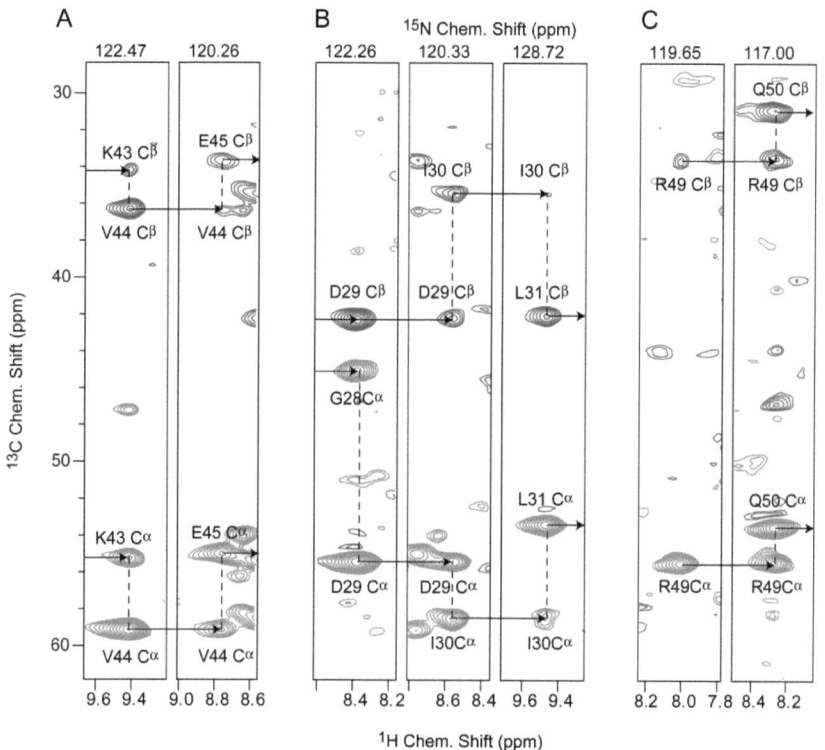

Figure 4.50. Assignment of residues E45 (A), I30 (B), and R49 (C) with an HNCACB experiment as examples of peak reassignment. The respective strip is plotted together with sequential contacts in order for an illustration of the assignment. The experiment was recorded at 600 MHz proton Larmor frequency, using 24 kHz MAS and an effective temperature of 22 °C. The sample contained 25 % protonation in exchangeable sites and 75 mM Cu(edta). Positive and negative contours are shown in green and blue, respectively.

New assignments

In order for an assignment of unassigned backbone resonances in the H/N-correlation, the temperature was varied in search for an optimal signal to noise using an HSQC pulse scheme. While at low temperature, the residues S36, T37, and N38 are visible, the majority of the formerly unassigned peaks shows up only at moderate or higher temperature. For G5 and T4, 22 °C seems an optimal effective temperature, since the peak intensity in an H/N-correlation goes down at even higher temperature. This behaviour might be due to a two-step chemical exchange process or an elevated

Methodological work on the SH3 domain

solvent exchange at higher temperature. Figure 4.51 shows parts of the H/N-correlation for three nominal temperatures, 257, 275, and 293 K, which corresponds to 3, 22, and 42 °C.

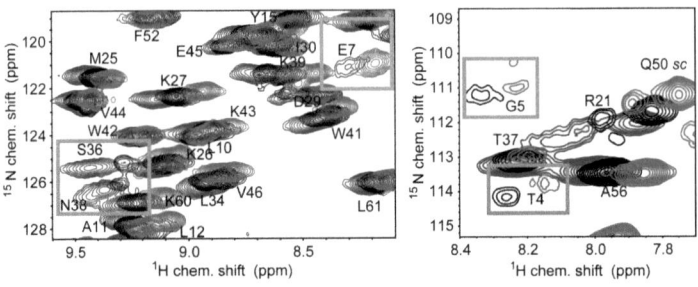

Figure 4.51. Comparison of the N/H correlation recorded for SH3 at different temperatures for the 10 % back-substituted sample. Colours refer to 257 K (blue), 275 K (black), and 293 K (red) nominal temperature, according to 3 °C, 22 °C, and 42 °C. Resonances which yield higher intensities at either elevated or lower temperature are highlighted by the surrounding grey rectangles. All spectra were recorded, processed and plotted using identical parameters (see Experimental Section).

Most assignments of the SH3 domain could be checked and, in some cases, reassigned. For the majority of the formerly *unassigned* resonances, no sufficient signal to noise is obtainable in an HNCACB experiment even after a manifold variation and optimization of the experiment. This refers to development and improvements of the pulse sequences, decoupling, Paramagnetic Relaxation Enhancement (PRE), and the protonation degree of exchangeable sites. However, the interplay of HNCA, HNCO, HNCACO and the TROSY-versions of these experiments give sufficient information about the resonances for an assignment based on chemical shift comparison with the solution state shifts.[177] Figure 4.52 displays the backbone walk for the N-terminal residues in the C^{α} and CO chemical shifts. The depicted assignments match the results of the RFDR-experiment represented in Figure 4.47.

Methodological work on the SH3 domain

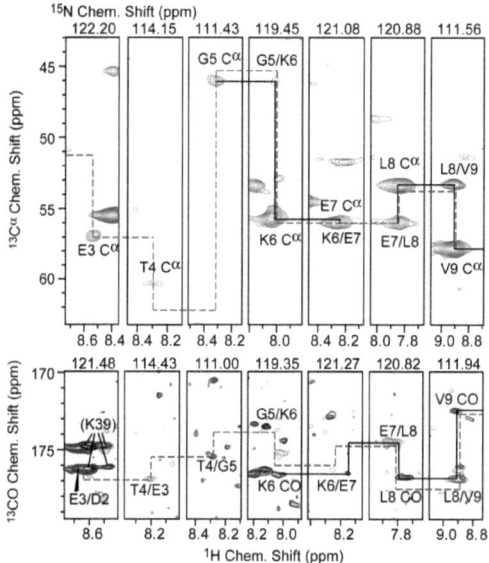

Figure 4.52. Assignment of the N-terminus based on the comparison with solution state shifts. The combination of HNCA spectra (green), HNCO (light blue), and HNCACO (dark blue and violet) enable a comparative assignment. The upper row displays the ^{15}N chemical shifts of each strip. An addition to an incomplete backbone walk shown in black, the solution state chemical shifts are represented by the red dashed lines, giving an imaginary "solution" backbone walk. These resonances were used to fit the assignment of the solid state. As a second parameter, the amide resonances (^{1}H and ^{15}N) were tried to match the solution values (see below).

The carbon chemical shifts are expected to move less than the amide moiety resonances, since different H-bond characteristics altered by crystal crystal contacts easily result in chemical shift changes of the involved nuclei. The great majority of the resonances, however, show largely similar chemical shift values also for amide nitrogens and protons.[173,177] Thus, a chemical shift comparison was pursued also for the N-terminal amide resonances. Figure 4.53 shows the H/N-correlation assigned with previous assignments[177] (black), new assignments according to HNCACB and HNCA spectra (blue) and the side chain assignments obtained by the combination of these experiments with ^{13}C direct excitation experiments (green, see below). In addition, solution state chemical shifts of the missing resonances are shown with red crosses and labels. Except for E7, all unassigned resonances were found to match the closest solution resonances in the H/N-correlation also in the carbon chemical shifts of the backbone.

Methodological work on the SH3 domain

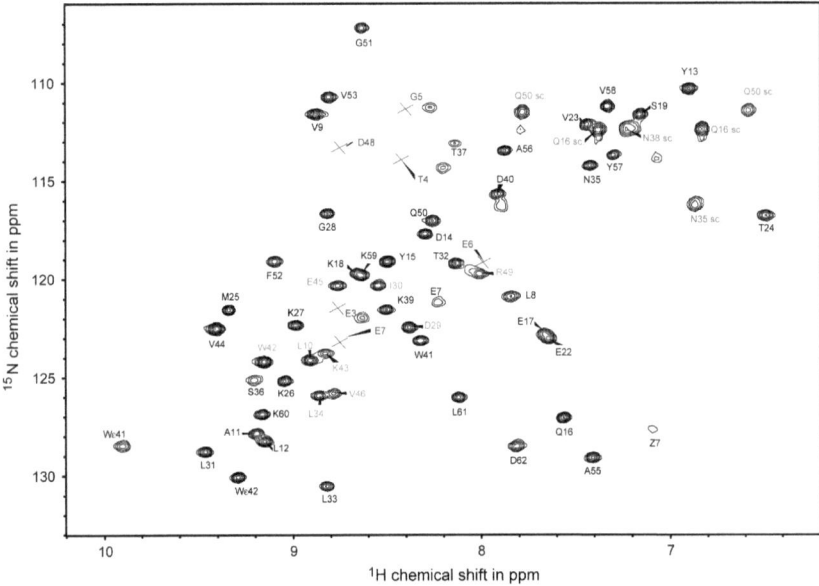

Figure 4.53. Comparative assignment of the N-terminal resonances concerning proton and ^{15}N chemical shifts. The solution state resonances are depicted with red crosses. Reassigned residues due to an improved resolution of assignment experiments as described in Chapter 4.2 are depicted in blue. Residue labels written in green denote side chain correlations assigned by the combination of solution state experiments, whole side chain experiments and direct ^{13}C excitation experiments.

The amide moiety of glutamine and asparagines side chains is detectable in the H/N-correlation due to two ^{1}H resonances with a degenerate ^{15}N chemical shift. As probed for the SH3 domain of α-spectrin, out-and-back experiments optimized for a backbone-assignment also lead to reliable side chain amide assignments in most cases. As described in Chapter 4.4 , direct ^{13}C excitation experiments were used in addition to obtain more complete information about the side chain amides. The derived assignments are depicted in Figure 4.53 (green). The downfield resonances of N38$_{sc.}$ and N35$_{sc.}$ could not be assigned unambiguously, neither was the side chain of N47. By exclusion, however, it is probable that the remaining pair of resonances next to Y57 with a ^{15}N chemical shift of 113.5 ppm are the N47 side chain correlations.

The chemical shifts of the backbone amide resonances of R21 have been obtained as 8.1 and 112.2 ppm for the ^{1}H and ^{15}N dimension, respectively.[177] Since the Q50$_{sc.}$ resonance has quite

Methodological work on the SH3 domain

similar shifts, the respective peak in the H/N-correlation has been interpreted as R21 in previous studies.[59] The side chain amide of Q50 is not involved in any kind of H-bonds and sticks to the solvent. Accordingly, a high flexibility is observed. This gives rise to significant outliers in dynamic studies.[128] The assignment of the side chain amide moiety of Q50 is shown in Figure 4.54. The assignment is possible only by a combination of different experiments. The signals in the HNCACB experiment are weak, but can be supported in part by the direct ^{13}C excitation CACOH experiment (vide ultra). The degeneracy of the chemical shifts of R49 C^β and Q50 C^γ can be enlightened by an additional strip through the backbone amide group of G51 in the whole side chain correlation.

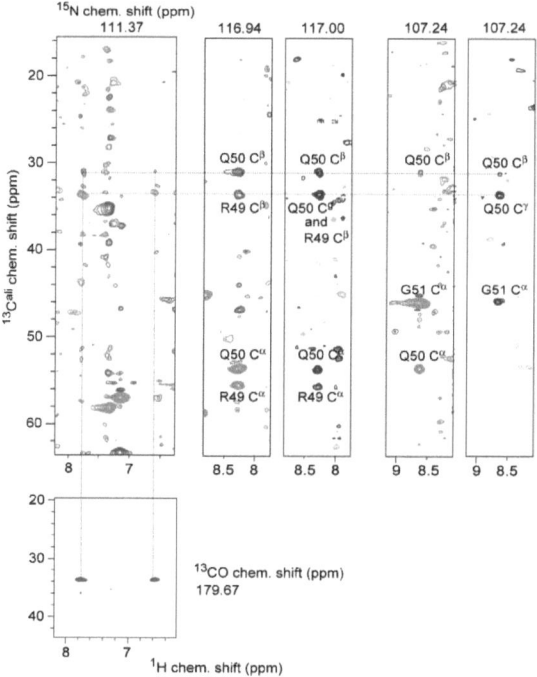

Figure 4.54. Assignment of the side chain amide resonances of Q50. The HNCACB experiment shows only weak signal intensity for the side chain amide correlation (left) even for a several fold optimized PRE and proton content of the exchangeable sites. The additional experiment with direct ^{13}C excitation (pink) supports the C^γ chemical shifts. Since the C^γ chemical shift coincides with R49 C^β, the strip through Q50 in the whole side chain experiment (blue) does not seem to show C^γ (3rd strip) That this shift indeed coincides with C^γ is confirmed by the strip through G51 of the whole side chain experiment (5th strip), where appearance of

R49 C^β is not expected. For completeness, the HNCACB strip through G51 is depicted in addition (4th strip).

Unfortunately, we do not observe a different peak that we can clearly assign R21. There is a small chance that the resonances coincide with $Q50_{sc.}$ (E) (downfield ^{1}H resonance). More likely, however, R21 undergoes slow chemical exchange und is not visible in the experiments due to fast relaxation.

The additional information about side chain mobility are of great use for studies concerned about intermediate and slow dynamics. For the respective groups in the SH3 domain, differential relaxation is represented in Figure 4.55. The data was obtained by the experiments described in Chapter 4.3 (see Figure 4.24). For the side chain amide groups, differential relaxation rates are less understood due to the moiety's constitution deviating from that of the protein backbone. Particularly slow motion seems to apply to N47. This residue is located in the distal loop, and in our studies, its backbone amide has not provided clearly detectable/assignable resonances so far. In the crystal structure, N47 is refined with two conformations.[60] $N38_{sc.}$, which seems to undergo dynamics on a slow timescale only at higher temperature, takes part in two H-bonds, namely to T32-OH and I30-CO for the Z- and E-proton, respectively. Q50, on the other hand, is directed to the bulk solvent and undergoes fast motion due to low restriction. The same is true for Q16.

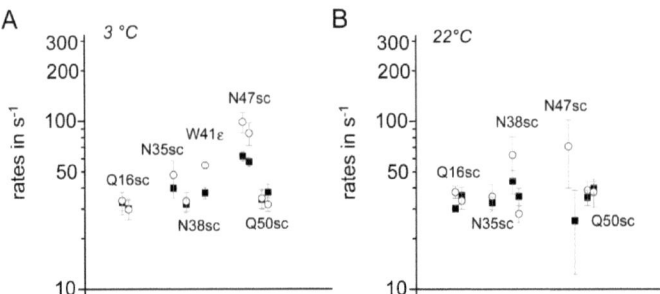

Figure 4.55. Transverse relaxation rates R_2 of the narrow $H^\alpha N^-$ and the broad component $H^\beta N^-$ in an H/N-correlation for two different temperatures, 3 °C (**A**) and 22 °C (**B**). The experiments are described in Chapter 4.3, Figure 4.24. Not all moieties could be characterized due to insufficient signal to noise. The dynamics displayed by side chains reflect their function in the protein structure. The left and right value for each amide group in the Figure displays low- and highfield component, respectively.

Methodological work on the SH3 domain

The amide moieties of asparagine and glutamine side chains are prone to showing extraordinary mobility. In addition, the side chain amide protons may exchange with water quite easily (for an outward position) or rather not (for an involvement in a salt bridge in the protein core), dependent on their function. If the respective side chain is directed to the bulk solvent, it can (dependent on the activation barrier) undergo rotational exchange or a direct exchange of protons between the E and the Z position. A significant exchange process will lead to cross peaks in experiments correlating the two states, as shown in Figure 4.56 A. In general, these dynamics can be considered for an understanding of protein constitution in a biological context. Accordingly, for methodological approaches, these groups are of special interest.[174] The assignment of the resonances in the SH3 domain has provided a quantification and understanding of the exchange process of $N35_{sc.}$ (as shown for the dipolar CODEX experiment[174] in Figure 4.56 B).

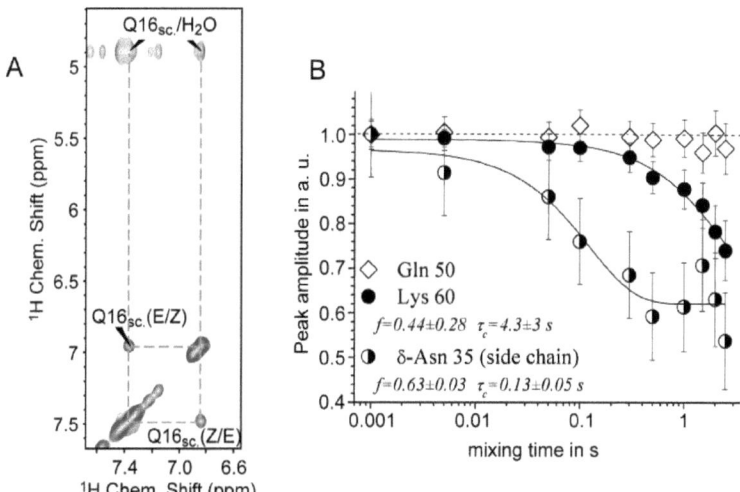

Figure 4.56. Dynamics of side chain amides. (**A**) Chemical exchange of the side chain amide moiety of Q16. The experiment was recorded using a ^{1}H-NOESY-^{1}H/^{15}N -type experiment similar to that in Figure 4.40 B with a mixing time of 150 ms without use of RFDR pulses. The exchange with water protons is also visible. (**B**) Side chain dynamics observed by Dipolar CODEX spectroscopy.[174] Particularly strong mobility is suggested for $N35_{sc.}$ by a decreasing peak amplitude for an increasing delay between dipolar encoding and decoding.

5 PROTON DETECTED SOLID STATE NMR APPLIED TO $A\beta^{1-40}$

(Work in this chapter was published in publication #10 on page 183)

As described in Chapter 2.2, the 40 residue peptide $A\beta^{1-40}$ is produced after cleavage of the Amyloid Precursor Protein (APP) by the β-secretase BACE and γ-secretase in the brain of Alzheimer's patients. The peptide tends to form oligomers and high-molecular weight polymers by aggregation. In contrast to these fibrils, no secondary structure is adopted in solution. This is shown in Figure 5.1, representing the H/N-correlation of $A\beta^{1-40}$ dissolved in DMSO.

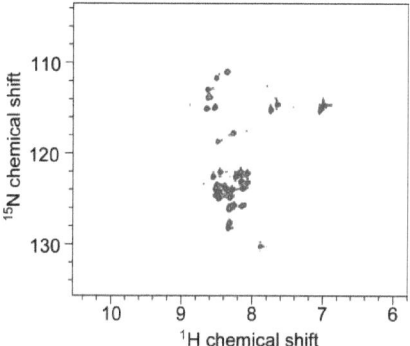

Figure 5.1. Solution state H/N-correlation of $A\beta^{1-40}$ in DMSO. The random coil structure is recognizable by the clustered amide resonances between 8 and 8.5 ppm and around 125 ppm in the ^{1}H and ^{15}N dimension. The only residues with a different chemical shift (downfield shifted in the ^{15}N dimension) pair are the glycines.

Little of this behaviour is maintained upon fibrillization. The resulting spectra are dominated by large downfield shifts in particular in the proton dimension, which is indicative for a β-sheet structure of the fibrils. For an assignment and an eventual structure calculation, this is advantageous since peak overlap tends to be reduced for β-sheet structures. However, in contrast to solution, a massive heterogeneity complicates standard experiments. This is due to peak overlap and a largely reduced peak height of the same integral peak intensity. Thus, in order for any successive work on the peptide, preparation conditions providing an optimal sample homogeneity are required.

Application to Aβ$^{1-40}$

5.1 Sample preparation and spectral quality

The sample preparation was done by Uwe Fink and Murali Dasari. For completeness, the most important samples on the way to spectroscopically valuable conditions are listed in Table 5.1.

Sample	H$_2$O/D$_2$O	[[Cu(edta)]$^{2-}$]	other conditions
A	15 %	75 mM	doping upon fibrillization, 40 % glycerol, 20 mM Tris, 8 mg
B	10 %	100 mM	doping only after formation of fibrils, 40 % glycerol, ca. 8 mg
C	30 %	75 mM	doping upon fibrillization, seeded, ca. 20 mg
D	50 %	75 mM	doping upon fibrillization, seeded, 20 mM Tris, 20 mg
E	25 %	75 mM	doping upon fibrillization, seeded, 25 mM NaCl, 45 mg

Table 5.1. Preparation conditions for the most important subset of samples. All samples were filled into 3.2 mm rotors. Sample amounts are approximate.

To achieve high signal to noise, paramagnetic doping as described in Chapter 3 was sought for the Aβ$^{1-40}$ fibrils. 10 % proton back-exchange in the absence of PRE yielded a longitudinal relaxation time T_1 of 1.3 s. While an HSQC with decent signal to noise (~8:1) could be recorded within 3.5 d for this non-doped sample, the spectra shown in Figure 5.2 were recorded in around 12 h. These spectra were recorded for samples A and B, both containing Cu(edta). Two ways of sample doping were implemented, (A) fibrillization in the presence of Cu(edta) and (B) washing of preformed fibrils in a Cu-edta containing buffer (B). The signal to noise of the sample A is approximately 10:1. Here, a protonation level of 15 % and a concentration of 100 mM Cu(edta) was employed. The spectrum in Figure 5 B was recorded for a 10 % back-exchanged sample with 75 mM Cu(edta) (sample B). This latter spectrum yields only half the signal to noise in comparison to the spectrum in A.

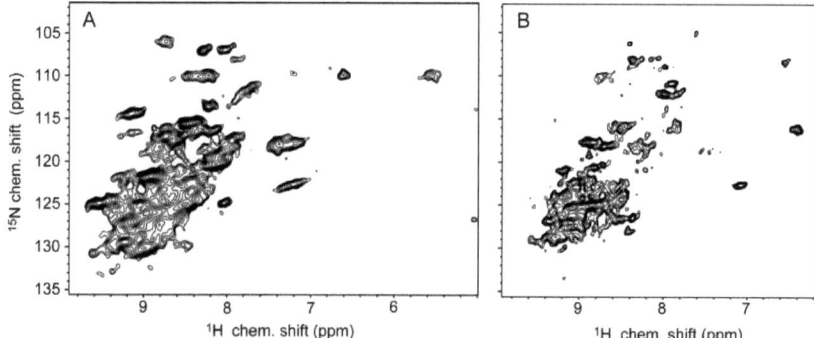

Figure 5.2. Preparation of fibrils doped with Cu(edta) added either upon fibrillization (**A**) or after fibrillization (**B**), using samples A and B of Table 5.1, respectively. Due to the heterogeneity of the sample no specific differences can be observed. In order for the most representative appearance of the samples, the spectra were processed using QSINE apodization. The spectrum in A has a proton content of 15 % in exchangeable sites and displays a better signal to noise (~2 fold), both experiments were recorded in approximately 12 h. The two samples were prepared without seeding.

We found a decrease of longitudinal T_1 times from 1.3 s without doping to 0.3 s (0.5 s) upon doping with 100 mM (75 mM) Cu(edta)) for the preparations (A) and (B), respectively.

In order to increase the homogeneity of the sample, fibrils were generated by seeding (see Chapter 2.2). The gain in sample quality by a nucleation is depicted in Figure 5.3. In these preparations (samples C and D, respectively), the amount of protons was increased while the level of PRE was maintained. The H/N-correlation of sample C in Table 5.1 with a proton content of 30 % (depicted in Figure 5.3 B) yields a better signal to noise also in triple resonance experiments based on N/C INEPTs (see section 4.1). Despite an increasing signal to noise in 2D H/N-correlations, a proton content of 50 % (sample D in Table 5.1) turned out to induce inconvenient relaxation properties for assignment experiments.

Application to Aβ$^{1-40}$

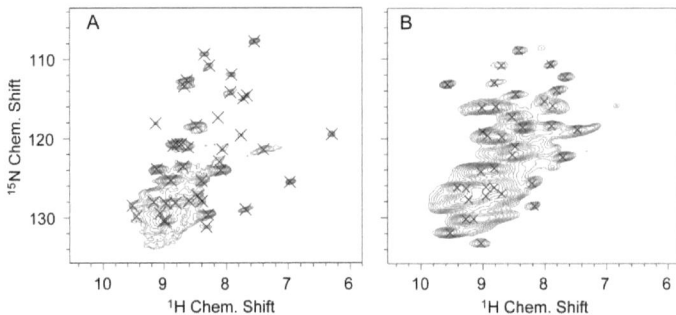

Figure 5.3. Spectra obtained upon use of seeded fibrillization for two different levels of proton back-exchange. The sample in (A) contained 30 % protons in exchangeable sites (sample C in Table 5.1), the one in B contained 50 % (sample D). The higher degree of protonation resulted in good signal to noise for H/N-correlations but turned out to be inconvenient for backbone assignment experiments based on N/C INEPTs (see section 5.2).

In comparison to protonated samples, isotope induced shift changes result from deuteration (ca. 0.35 ppm/proton).[33] These effects are particularly significant for methylene or methyl groups due to two-fold substitution of bonded atoms.

Application to Aβ[1-40]

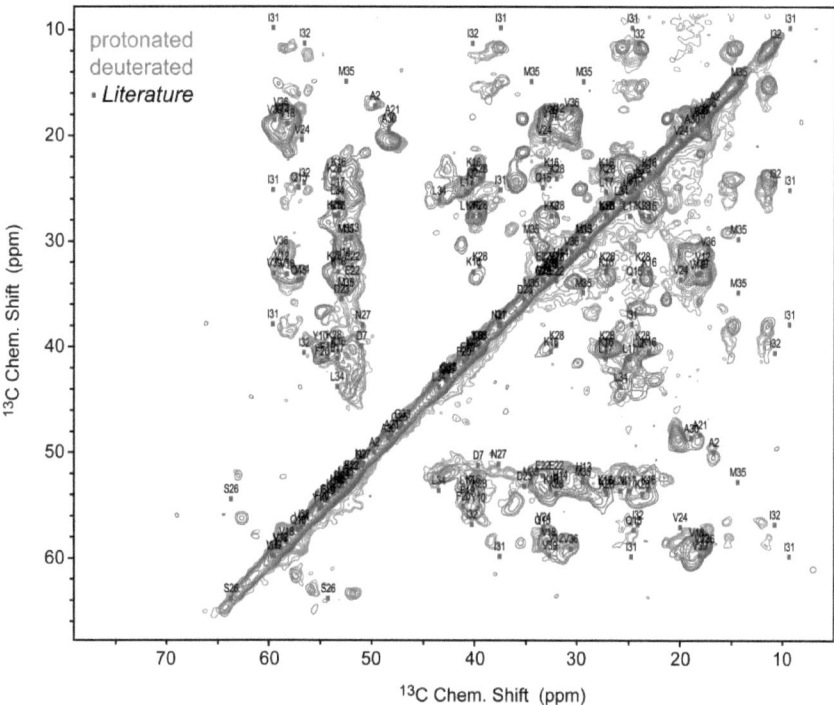

Figure 5.4. Comparison of C,C-correlations obtained for deuterated (25 % ^{1}H proton back-exchanged at labile sites and doped with 75 mM Cu(edta), blue) and for uniformly protonated Aβ[1-40] fibrils (green). In addition, chemical shift values taken from Paravastu et al.[90] are depicted using red squares. Mixing among carbon spins was achieved using TOBSY (15 ms) and PDSD mixing (50 ms) for the deuterated and protonated sample, respectively. For the acquisition of the first correlation, ^{2}D-decoupling was applied during indirect and direct detection.

5.2 Backbone assignment

For systems that show increased relaxation due to higher concentration of protons (see Chapter 4.1) or PRE, INEPT steps may lead to severe relaxation losses in comparison to CP based experiments. In this work, higher protonation levels were used in particular in the course of the Aβ[1-40] assignment. Additionally, we observed this system to yield significantly higher relaxation rates

Application to Aβ$^{1-40}$

than the SH3 domain even for an identical content of proton. This is maybe due to dynamics or different behaviour towards Cu^{II}-PRE (see Chapter 3). In these cases, an exchange of INEPT magnetization transfer by Cross Polarization (CP) steps turned out to be a reasonable expansion of the solid state out-and-back triple resonance experiments. Use of CP provides a reduction of transverse relaxation during ^{1}H/^{15}N magnetization transfer in comparison to J-based sequences, because the CP step (~250 µs) is usually much shorter than the magnetization transfer by INEPT (~5 ms) (see Chapter 1). For a sample in which exchangeable protons are 30 % back-substituted, we observed a more than two fold increase in sensitivity. For 50 % back-exchange, INEPT-based spectra did not yield more than a tenth of the respective CP experiment, however, this comparison does not consider that just a subset of the protein residues (especially flexible residues) are detected (compare Chapter 4.1). Figure 5.4 shows the first increment of an HSQC of Aβ$^{1-40}$ fibrils with 30% and 50 % protonation in exchangeable sites (samples C and D in Table 5.1), respectively, in comparison to that of an H/N correlation recorded with the help of CP.

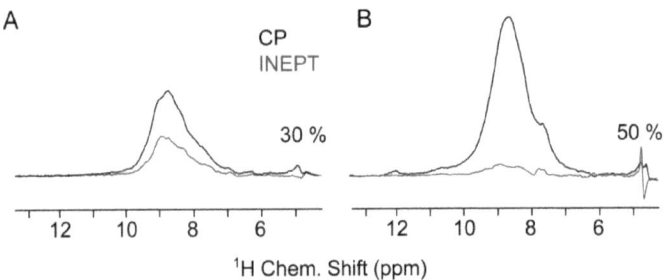

Figure 5.4. Comparison of INEPT and CP for a H/N-correlation for 30 % (A) and 50 % back-exchange at H^N sites (B) (samples C and D in Table 5.1). Black spectra were recorded according to Chevelkov et al.[59] using a ramped CP, while the red spectra depict the first increment of an HSQC experiment. All experiments were recorded under identical conditions using 128 scans. Slightly different amounts of protein might have to be taken into account in case of a comparison between A and B.

Implementation of H/N-CPs into solution state out-and-back experiments is straight forward replacing the respective INEPT steps against CPs. A ramped CP was used to ensure a higher resistance against experimental instabilities in the course of longer experiments. Additional changes exclusively apply to the phase cycle. Additional ^{1}H decoupling during transverse ^{13}C magnetization did not yield any improvement. Figure 5.5 shows the pulse schemes for HNCA and HNCACB

Application to Aβ[1-40]

experiments relying on CP magnetization transfer from ^{1}H to ^{15}N. To yield correlations between covalently bonded atoms only, the ^{15}N/^{13}C transfer was performed as INEPT in all cases. For proteins of low proton concentration, this feature performed better in terms of signal to noise as well. For higher protonation levels, CP (or other dipolar based techniques like REDOR), however, are an alternative.

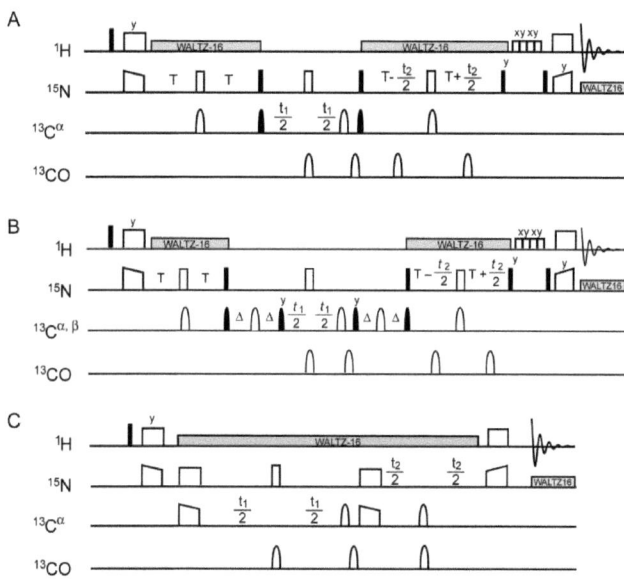

Figure 5.5. Pulse schemes for CP based out-and-back experiments. HNCA (**A**) (interchange of CO and C$^\alpha$ pulses yields an HNCO scheme) and HNCACB (**B**). Alternatively, decoupling can be used during ^{13}C transverse magnetization as well. Open bars of reduced height refer to water purge pulses of alternating phase.[162] **C**) depicts a complete exchange of J-based against dipolar based transfer. For SH3, intensities were found to be lower than A) and B) for low proton content (10 % back-exchange) but are an alternative if higher protonation levels are used.

At the same time, the implementation of CP transfers holds the advantage that no antiphase ^{1}H/^{15}N coherence has to be refocused. For samples in which ^{15}N transverse magnetization relaxes quickly due to ^{1}H/^{15}N dipolar interactions, the immediate application of ^{1}H heteronuclear decoupling hence increases the sensitivity additionally. This is favourable, since increased transverse relaxation in ^{1}H (during the INEPT) and in ^{15}N naturally appear in the same samples.

Application to Aβ$^{1-40}$

Figure 5.6 displays a comparison between the first increment of an INEPT and a CP based HNCO experiment with and without decoupling for an Aβ$^{1-40}$ sample containing 30 % back-substituted proton content at exchangeable sites. Obviously, decoupling is even more effective in combination with CP.

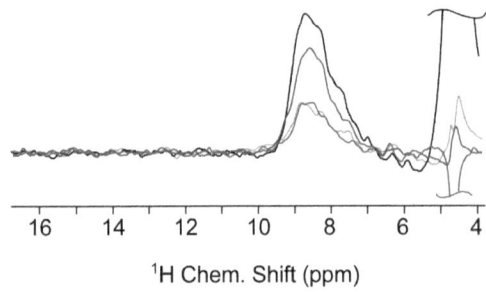

^{1}H Chem. Shift (ppm)

Figure 5.6. Comparison of the performance of heteronuclear ^{1}H-decoupling during transverse ^{15}N magnetization in case of CP transfers (green: without, black: with decoupling) and INEPT transfer (without: blue, with decoupling: red) in HNCO experiments of Aβ$^{1-40}$. The spectra were recorded under identical conditions on a 30 % ^{1}H back-exchanged Aβ$^{1-40}$ sample with 2048 scans each (sample C in Table 5.1).

Using INEPTs, decoupling may increase the sensitivity of the triple resonance experiment by a factor of 2.1. Due to the absence of a H_zN_x refocusing delay (5.5 ms), the CP based experiment was found to gain by decoupling by a factor of 2.8. Due to these effects, the ratio between decoupled CP and INEPT based experiments is approximately 1.3 for 30 % H^N protonation. For 50 % ^{1}H back-substitution, INEPT based out-and-back experiments perform even worse in respect to CP-based ones.

In the case without decoupling, ^{1}H/^{15}N-^{1}J-couplings were refocused by 180° composite pulse concomitant to the off-resonance ^{13}C pulses. For the other case, decoupling was found to perform best using WALTZ-16^{132}, employing ~85 μs pulse duration and an approximate field strength of 8 kHz. We did not use decoupling of higher power in the course of the ^{15}N transverse magnetization periods (48 ms per experiment) in order not to compromise the sample integrity.

A second benefit of the absence of antiphase ^{1}H/^{15}N coherence is the possibility of a largely improved water suppression. The improved spectral quality can be appreciated in Figure 5.7, which

Application to Aβ$^{1-40}$

shows a comparison between first slices of HNCO spectra of Aβ$^{1-40}$ fibrils with 30% back-substitution of exchangeable protons using INEPT transfer with a water purge pulse of 1 ms duration and 60 kHz power (red) and CP with four 25 ms water dephasing pulses of 10 kHz field strength adapted from the MISSISSIPPI sequence published by Zhou et al. for a facilitated water suppression with protonated (natural abundance) samples (blue).[162] For comparison, the Figure depicts a CP-based HNCO experiment without any water suppression in addition (black).

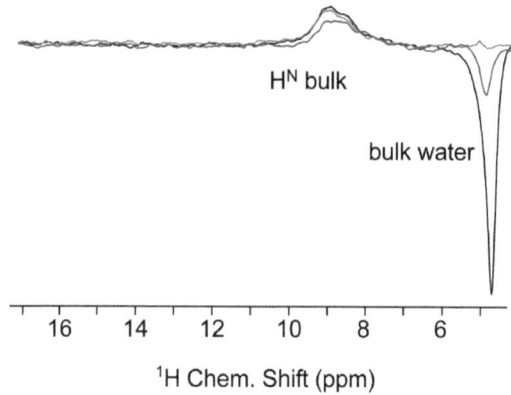

Figure 5.7. Comparison of CP based HNCO without water suppression (black) to the water suppression achieved in an INEPT based HNCO using a water purge pulse (red) and water suppression achieved by use of long purge pulses during ^{15}N longitudinal magnetization of a CP-based HNCO experiment (blue). Almost no water signal is observed even in the 3D spectra after application of four 25 ms pulses of 10 kHz with alternating phase on ^{1}H.[162] Spectra were recorded using 512 scans each (sample C in Table 5.1).

The use and performance of triple resonance assignment experiments in case of Aβ$^{1-40}$ is demonstrated in Figure 5.8. The HNCA and HNCACB experiments were recorded with the CP-based pulse schemes given in Figure 5.5. Although the HNCA spectra yield all resonances with a reasonable signal to noise of approximately 6:1 after 2.5 d, the assignment process is hampered by an overlap of C$^\alpha$ chemical shifts. This degeneracy is partly due to unrefocused C$^\alpha$/C$^\beta$ couplings and consequently broad lines. On the other hand, the sample inhomogeneity mentioned in Chapter 5.1 results in broad resonances. The use of an HNCACB can potentially resolve ambiguities in the C$^\alpha$ resonances, however, due to a fast signal decay in the course of the J-coupling evolution between C$^\alpha$ and C$^\beta$, resonances are weak or absent in many cases. This is true especially for C$^\beta$ resonances (except for alanine residues) because of the evolution of homonuclear couplings to C$^\gamma$. As an additional experiment yielding C$^\alpha$, C$^\beta$, and (partly) C$^\gamma$ resonances, a correlation recorded with the PAIN

Application to Aβ$^{1-40}$

pulse scheme^{51} helps identifying resonances in case of a low signal to noise (depicted in Figure 5.8 in columns annotated C).

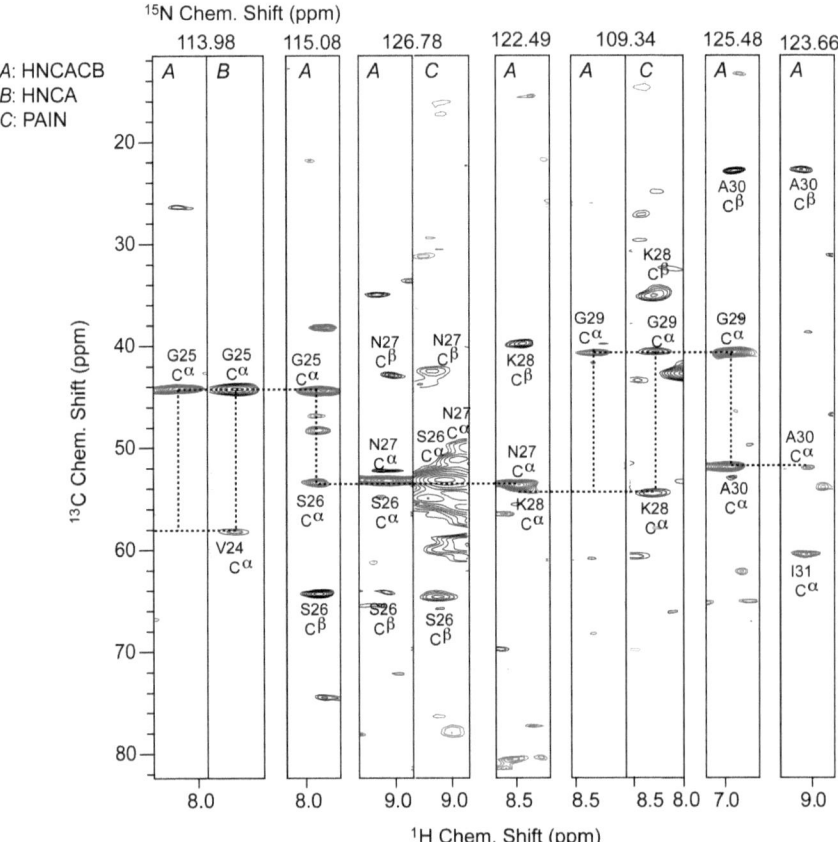

Figure 5.8. Visualization of the sequential backbone assignment of Aβ$^{1-40}$ using (A) HNCACB, (B) HNCA and (C) PAIN (Ref.) experiments represented for residues G25 to I31. Due to unambiguities, the assignment by Paravastu^{90} was used as an orientation. The experiments were conducted using a sample with 25 % H$_2$O and 75 mM Cu(edta) (sample E in Table 5.1).

Although the sample fibrillized in 30 % H$_2$O (see Figure 5.3 B) did not yield enough signal to noise for an HNCACB experiment, the HNCA could be used as an additional help for the assignment of backbone resonances of the peptide. This is due to an improved sample homogeneity in

Application to Aβ$^{1-40}$

comparison to the sample used in the other experiments. Figure 5.9 compares the respective backbone walk for residues V36 to V40, extracted from the HNCA and the HNCACB of the 30 % and the 25 % H$_2$O sample in A and B, respectively.

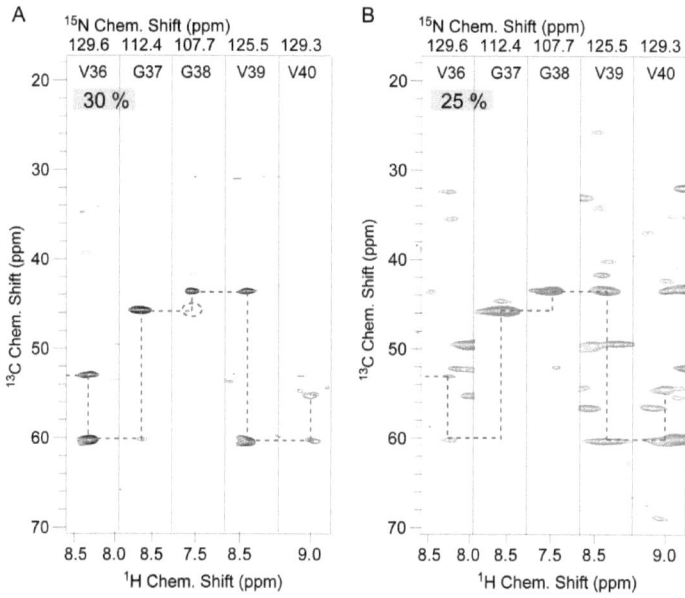

Figure 5.9. Representation of the improved sample homogeneity of the 30 % H$_2$O sample (sample C in Table 5.1) shown in (**A**) in comparison to the sample used in the remainder of the experiments (sample E in Table 5.1) shown in (**B**), which yields a better signal to noise. Supposedly due to a lower amount of material in the sample, no HNCACB experiment was obtainable for sample C.

For both latter samples, HNCO experiments could be recorded in a straight forward way. Figure 5.10 shows a representative part of an exemplary HNCO experiment, recorded with a sample having a proton content of 30 % at labile sites and 75 mM Cu(edta) (sample C in Table 5.1). The excerpt displays residues M35 to V39.

Application to Aβ$^{1-40}$

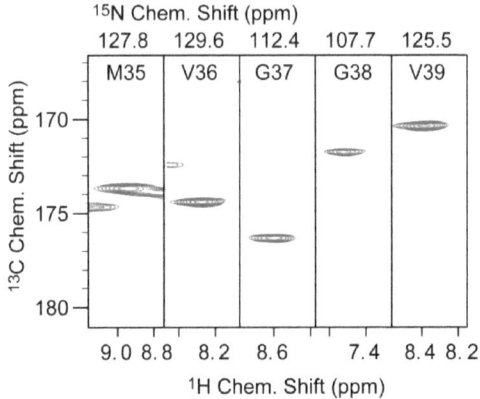

Figure 5.10. Example of an HNCO experiment, recorded at 600 MHz Larmor frequency, 24 kHz MAS and a temperature of 22 °C. The strips show a representative part of the residues for the case of a 30 % proton back-substituted sample (sample C in Table 5.1). The spectrum was recorded with the pulse sequence shown in Figure 5.5 A, exchanging C^α and CO pulses.

As described in Chapter 4.5, RFDR mixing among protons has been shown to yield correlations between dipolar coupled protons. This can be used for elucidation of distances between assigned H/N resonance pairs. In addition, for non-assigned resonances, the correlations can be used to probe proximities within the primary sequence of the protein. Sequential contacts yield relatively strong cross peaks, as can be seen from the spectra in Chapter 4.5. In case of Aβ$^{1-40}$, the dispersion of ^{15}N and especially ^{1}H resonances is comparably weak. Consequently, the gain of H/N based spatial correlations is limited by the overlap of signals in the respective dimensions. Figures 5.11 A and B display the general strategy of the N- and H-resolved amide correlations, respectively, for the example of residue A21. Due to only one resolved shift of the distant amide group, one experiment alone is not enough in order to know which cross peak belongs to which of the diagonal peaks with degenerate shift. Combining the two experiments for one residue allows to exclude false cross peaks. This is described in more detail in Chapter 5.4. Unfortunately, the information obtained by these experiments was insufficient for an unambiguous assignment in the case depicted in Figure 5.11.

Application to Aβ[1-40]

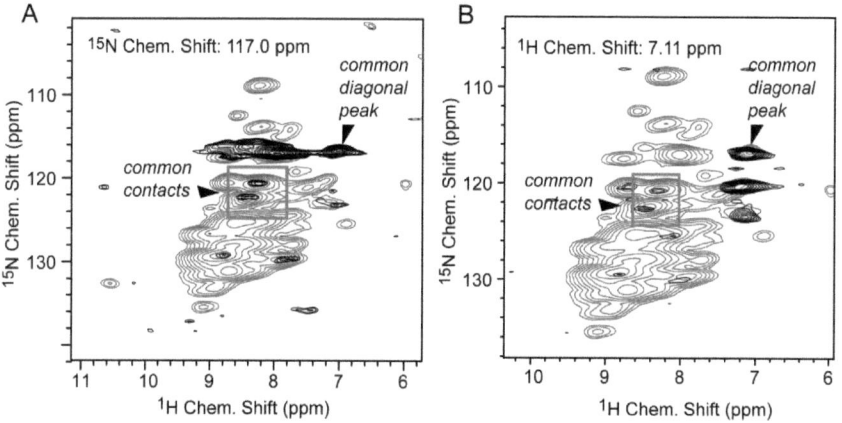

Figure 5.11. Experiments for the determination of amide moieties in close spatial proximity (black) overlaid with an H/N correlation with identical procession (red). **A)** The ^{15}N chemical shift of the distant amide group is resolved in F1, **B)** the respective ^{1}H chemical shift is resolved in the indirect dimension additional to the H/N-correlation. Displaying the H/N-slice for an arbitrary shift of the F1 dimension yields H/N-coordinates of all connected amides. By combination of the two experiments for one common cross peak, ambiguous contacts can be dissected. The employed sample is listed as "E" in Table 5.1. The pulse programs employed for spectra A and B are depicted in Figure 4.40 A and B, respectively.

Figure 5.12 shows the H-resolved experiment for spatially close amide moieties in case of G38, where a strong cross peak to G37 arises. In cases like this, the correlation yielding spatial contacts could be used for a confirmation of the assignments made by HNCA and HNCACB type experiments. The traditional purpose of experiments like these ones, i. e. providing distance restraints for structure calculation, however, could not be approached due to an incomplete assignment of the Aβ[1-40] fibrils.

The HN-correlation of Figure 5.12 displays the preliminary assignment based on the various experiments described before (HNCA, HNCACB, PAIN, RFDR-experiments). A great part of the resonances, especially those lying in the region of unresolved chemical shifts, could not be assigned unambiguously. For this future aim, an improved sample preparation will help.

Application to Aβ$^{1-40}$

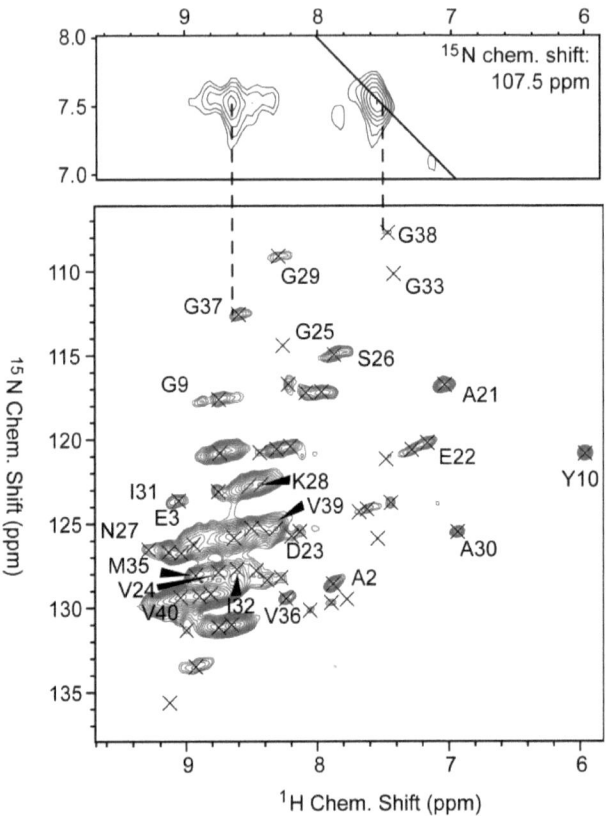

Figure 5.12. Representation of an unambiguous case in which the RFDR experiments could successfully be used for a confirmation of an assignment known before from HNCA/HNCACB experiments (top). The deduced assignment of the HN-correlation (bottom) is only preliminary, since a great part of the shifts could not be unambiguously dissected in the course of the backbone assignment.

5.3 Side chain spectroscopy

In the course of the acquisition of H/N-correlation experiments, histidine side chain signals could be observed for different Aβ$^{1-40}$ samples. The respective resonances of protonated imidazole groups are depicted for a H/N-correlation of a 30 % proton back-exchanged sample in Figure 5.13. In addition, an H/C-correlation (recorded by Vipin Agarval) is depicted for completeness (C).

Application to Aβ$^{1-40}$

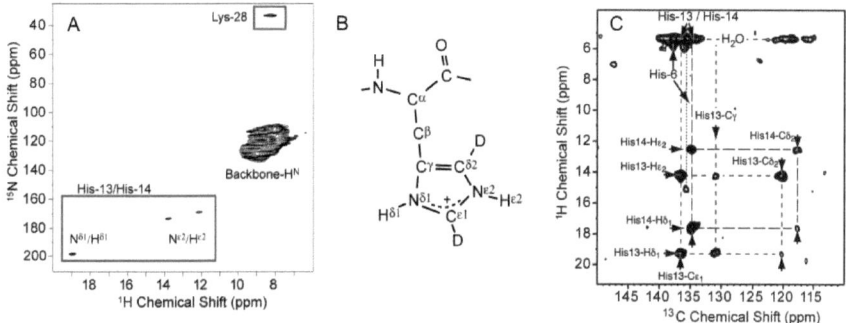

Figure 5.13. Histidine signals could be observed for experiments at low temperature (3 °C). A) H/N-correlation of a sample with 50 % protons at exchangeable sites showing side chain N/H correlation signals. The fourth peak is broadened to the noise level. B) and C) The spin systems as deduced from the chemical structure and as seen in H/C correlation experiments (recorded by Vipin Agarval). The spectra in A and B were recorded at 600 MHz ^{1}H Larmor frequency and 24 or 20 kHz MAS, respectively.

In future work, the assignment of these extremely shifted peaks will be aimed at, using the assigned backbone resonances described in Chapter 5.2. Further investigation into this direction is interesting as the extreme downfield shift of the His-resonances is supposedly due to Cu-binding, which may also play a role for the toxicity of Aβ$^{1-40}$ fibrils.[87]

CIDNP type effects observed for Aβ$^{1-40}$ fibrils at elevated temperatures

The presence of a histidine probably interacting with Cu causes a yet unclear observation. Recording experiments at higher temperatures, down-field shifted resonances were observable in the ^{1}H spectrum. These signals were phaseable in experiments with simple one-pulse excitation. Upon use of CP transfer, unphaseable spectra were obtained. This is documented in Figure 5.14, representing simple excitation experiments at 700 MHz ^{1}H Larmor frequency before and after temperature increase to 40 °C and the first slice of a CP based HSQC experiment.

Application to Aβ$^{1-40}$

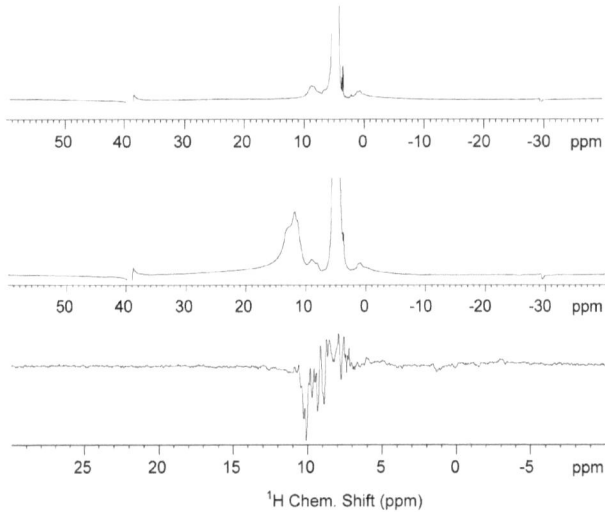

Figure 5.14. Example for down-field shifted signals arising upon heating the sample to ~40 °C. The unassigned signal has much higher intensity than the amide bulk signal and vanishes after several minutes. (top) ^{1}H spectrum at 30 °C. (center) Spectrum at 40 °C, both using simple single pulse excitation schemes. (bottom) Spectrum obtained at 40 °C upon use of CP for ^{15}N filtered proton 1D experiments. The spectra were recorded at 700 MHz ^{1}H Larmor frequency and 24 kHz MAS.

The unassigned signals reproducibly arose after heating of the sample to 40 °C and faded away after approximately 3 min. The observation was reproducible for two spectrometers, however, the chemical shift of the hump was altered to lower values at 600 MHz ^{1}H Larmor frequency. The little time span until the signal decayed was sufficient to record a preliminary 2D H/N-correlation. This spectrum is shown in Figure 5.15.

The origin of the signal arising downfield from the amide bulk at different chemical shift values is unclear and might be due to the influence of Cu bound to the histidine residues. A potential explanation might involve recombination of Cu radical pairs induced by thermal polarization in terms of the CIDNP effect.[178] The existence of unpaired electrons in Aβ fibrils has been supposed to be entangled with their toxicity.[179] Future investigation will aim at the elucidation of the observations made here.

Application to Aβ[1-40]

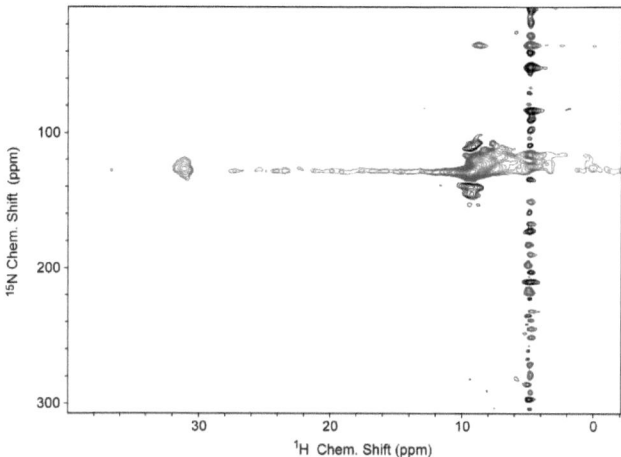

Figure 5.15. 2D H/N-correlation recorded after temperature increase from 30 to 40 °C. Due to a time slot of only some minutes, the ^{15}N dimension is severely truncated. The experiment was recorded at 600 MHz and 24 kHz MAS.

6 PROTON DETECTED SOLID STATE NMR APPLIED TO MEMBRANE PROTEINS

(Work in this chapter was also published in publication #10 on page 183)

The experiments developed in the course of the thesis have in part been applied to the membrane protein OmpG (Outer Membrane Protein G) already. The aim of a facilitated assignment and structure calculation by means of proton detected NMR, however, has been restricted by sample preparation in case of OmpG yet. Due to the long T_1 times found in extensively deuterated proteins, long recycle delays have to be used if no paramagnetic relaxation enhancement (PRE) is used. Respectively prepared samples in case of OmpG are planned to be produced in future studies.

Undoped samples of deuterated OmpG yield H/N-correlations qualitatively comparable to those of the SH3 domain. Supposedly due to a larger protein heterogeneity, ^{1}H (approximately 50 Hz) and ^{15}N linewidths (approximately 30 Hz) are slightly increased. Furthermore, significant overlap occurs due to the larger number of residues. Figure 6.3 shows an overlay of H/N-correlations recorded with scalar and with dipolar magnetization transfers.

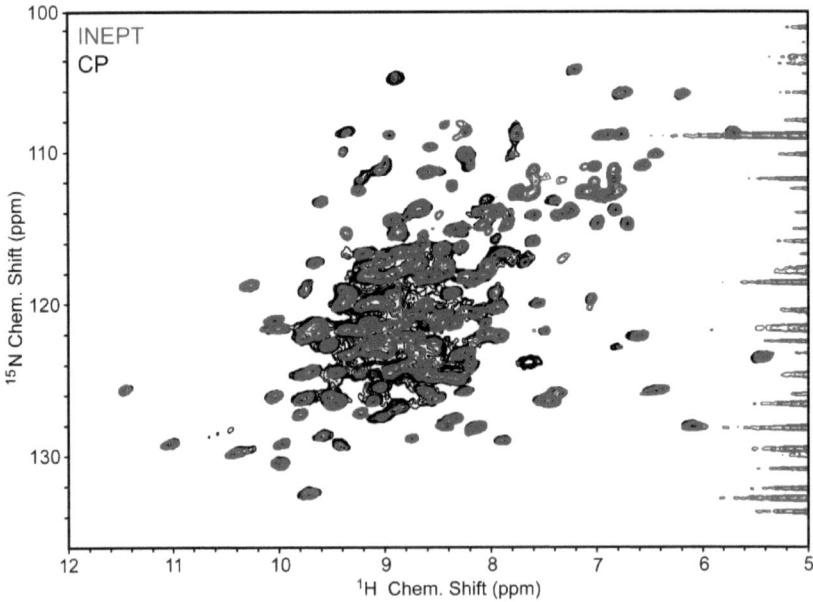

Application to membrane proteins

Figure 6.3. H/N-correlation of the Outer Membrane Protein OmpG, recorded using INEPT and CP for magnetization transfer. The sample was prepared using deuterated protein with a 25 % proton back-substitution of exchangeable sites and deuterated *E. coli* lipids. Acquisition of the spectra was performed at 850 MHz ^{1}H Larmor frequency, 20 kHz MAS and a temperature of approximately 25 °C. Apodization used a shifted sine-bell function in both dimensions.

The use of deuterated lipids was assumed to provide longer T_2 times due to reduced interactions between the nuclei in the protein with lipid protons. Figure 6.4 displays H/N-correlations recorded for samples crystallized with deuterated and protonated *E. coli* lipids. Minor differences are visible, which are supposedly due to a different chemical surrounding for residues close to the outside of the pore.

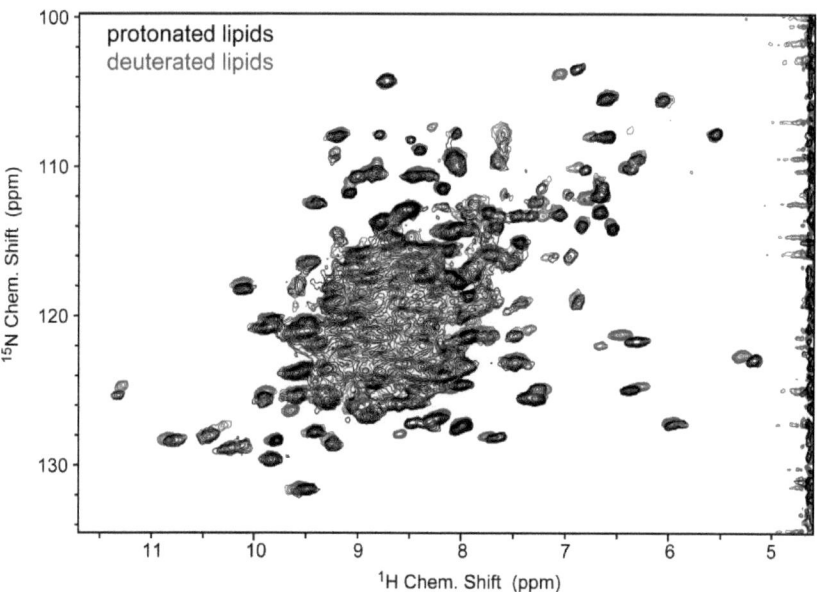

Figure 6.4. H/N-correlations recorded for samples crystallized with protonated (black) and deuterated (red) *E. coli* lipids. Both spectra were recorded identically using CP magnetization transfer at 850 MHz ^{1}H Larmor frequency and 20 kHz MAS at a temperature of approximately 25 °C.

In order to verify the assumption of longer T_2 values upon deuteration of the lipid environment, relaxation measurements were performed using relaxation delays refocusing evolution of chemical shift and *J*-couplings. The respective values measured for OmpG at 850 MHz ^{1}H Larmor fre-

Application to membrane proteins

quency with protonated and deuterated lipids are shown in table 6.1. The observed differences are significant, however, protein relaxation values do not seem to be influenced to a large extend by the lipid surrounding.

	^{13}CO T_2 (ms)	^{15}N T_2 (ms)	^{1}H T_2 (ms)	^{1}H T_1 (s)
Protonated lipids	5.7 ± 1.5	10.0 ± 1.0	6.8 ± 0.5	0.99 ± 0.04
Deuterated lipids	5.6 ± 1.5	11.1 ± 2.3	9.2 ± 1.0	1.16 ± 0.07

Table 6.1. T_2 for various nuclei and ^{1}H T_1 in samples containing protonated or deuterated *E. coli* lipids, respectively, with otherwise identical parameters. No ^{1}H decoupling was used for the relaxation measurements.

7 Discussion and Conclusions

Solid state NMR is an important tool for the elucidation of a vast range of biological processes in living organisms. This is particularly essential since the understanding of these processes provides the basis for improved preventive and therapeutic means against a broad range of diseases. For a deeper understanding of the structure and function of most biomolecules of contemporary interest, however, manifold obstacles, depending on the employed techniques, have yet to be overcome. For NMR techniques, these obstacles mainly base on the sensitivity and the accuracy that is achievable for a target molecule. Large scale protein production using recombinant expression has become a standard method in structural biology. Although host system cells like *E. coli* can be grown in isotope labelled media (e.g. D_2O in combination with specifically labelled carbon sources), mg-quantities of mammalian membrane proteins e.g. are difficult to express. In many other cases, the spectroscopic possibilities themselves (in terms of sensitivity and resolution) are insufficient even if the protein preparation provides sufficiently large amounts of protein.

The scope of the thesis focuses on the methodological improvement of techniques to allow proton detected solid state NMR. This technique is important for innumerable (biologically and pharmacologically important) proteins for which alternative techniques for elucidation of structure and function, like X-ray crystallography and solution NMR, are difficult. There are many examples for targets that are important for the understanding of a biological process but insoluble, difficult to crystallize or simply to large for these techniques. The spectral quality that is achievable in solid state NMR, however, is still far away from that obtained for NMR of small soluble proteins. Although the potential of NMR in the solid phase is promising, its applicability is still limited mainly due to technical complications.

The main obstacles for solid state NMR are limitations in signal to noise in combination with a low resolution that does not allow for clear differentiation between resonances of larger biomolecules. Both problems are largely alleviated by the results of this work. We chose the approach of an extensive proton dilution, which had been shown to dramatically increase spectral resolution. The sensitivity achieved traditionally had not been sufficient for most spectroscopic issues. This major problem could be alleviated by Paramagnetic Relaxation Enhancement, which allowed to enhance the former signal to noise by a factor of 3-4. This tool lead to an effective reduction of measurement time up to 15 fold and made it possible to combine an excellent resolution with a good signal

Discussion and conclusions

to noise. All of the follow-up strategies could not have been accessed without this signal enhancement. The applicability of PRE, however, is probably not ubiquitous. The SH3 domain is a comparably small protein domain. Here, the effect of paramagnetic doping could be shown to access all residues of this molecule. Although effective transfer of PRE also further into the protein by hydroxyl groups and water was confirmed, first trends of a differential approachability were observed (see Chapter 3.2). For applications on large molecules with an extensive hydrophobic core e. g., the question of how far a chelate complex may be away from a spin of interest without deterioration of PRE efficiency will have to be considered.

As a second important basis for a good sensitivity of an excessive proton dilution in addition to PRE, we improved the protonation level of exchangeable sites to an optimal value. The optimal concentration varies dependent on what type of experiment is to be performed. In general, however, in comparison to previous studies, another factor of 2-3 could be gained without losing spectral resolution.

Although these achievements are important independent results on their own, they also allowed for manifold improved spectroscopic strategies on perdeuterated protein samples. Among these techniques, the effective correlation of amide resonances with the carbon resonances of the backbone is the most important achievement. Similarly, a sequential assignment strategy based on solution state NMR techniques could be developed. In conclusion, we found two main advantages of this approach (in comparison to traditional solid state NMR): Firstly, the approach yields clean correlations between successive amino acids without non-sequential spatial contacts. Secondly, in the process of ^{13}C resonance mapping for sequential assignment, a clear separation of the ^{13}C resonances in two additional dimensions (by the amide nitrogen *and* proton) yields a very good resolution. This is in contrast to traditional experiments, where only ^{15}N chemical shift information is employed and high power decoupling determines recycle delays and resolution. In solution state, this assignment strategy has already been used for the assignment of proteins up to several tens of kDa. Although we found a different performance for different protein samples, it is expected that this highly unambiguous assignment strategy will greatly facilitate future solid state protein assignments.

Furthermore, the inclusion of proton chemical shifts turned out to also be viable for sidechain amide and sidechain ^{13}C resonance assignment. Due to larger $H^N/C^{ali.}$ distances, however, access of ^{13}C resonances in the sidechain had to be mainly based on direct excitation. This compromise is due to the difficulty of reaching side chain carbons far away from amide sites if exclusively pro-

Discussion and conclusions

tons are used for start magnetization. Consequently, the respective correlations suffer from a lower gyromagnetic ratio of the start nuclei and a longer recycle delay than in case of the backbone experiments. In conclusion, however, the direct excitation approach works surprisingly well, which dues to the high abundance of ^{13}C nuclei in comparison to the proton content. The resulting experiments, one of which corresponds to an HN-TOCSY in solution, are of enormous value for sidechain assignments of deuterated proteins. This approach appreciates the shorter recycle delay reduced by PRE in a particular way.

Taken together, the complete and accurate assignment of all resonances (^{1}H^{N}, ^{15}N, ^{13}C) of a solid state protein can be demonstrated using a single sample. Compared to the expensive and lengthy process of specific labelling, which has traditionally been used for many solid state studies, this is an important achievement, which will largely facilitate the applicability of future solid state NMR on proteins. An appreciable future extension of this approach might be the inclusion of other (dilute) protons than H^{N} for backbone torsion angles and a higher proton restraint density. This subject is a valuable prospective being worked at in our group.

Another result of the thesis is the observation and assignment of protein residues in dynamic regions undergoing slow motion. This achievement, which is based on the selection of spin states, has a particular value. The tool will be of great importance for future studies focusing on different kinds of protein functionality, since these are often accompanied by slow motional processes. With traditional techniques, the respective assignments are difficult or impossible to achieve and thus represent an important new focus for solid state NMR.

More as a side aspect, accurate structural restraints are obtained by measuring proton-proton distances. Still, this technique can only complement existing methods, since even long range proton contacts are sparse in comparison to e. g. ^{13}C/^{13}C distances. For the first time, however, these spectra have been obtained with an ultra-fine resolution and yield unambiguous correlations for up to ~7 Å, which is a much longer distance than those encountered in traditional solid (and even solution) state NMR.

The assignment strategies worked very well for the α-spectrin SH3 domain. The assignment of larger proteins (as shown in the last part of the manuscript), however, is not straight forward even with this methodology due to a lower number of molecules in the rotor and a decreased signal to noise. The work necessary for the completion of this task would span a time frame beyond the scope of this thesis. As a conclusion of this work, however, assignment and a successive structure determination of a membrane protein like OmpG without use of site specific labelling schemes is

Discussion and conclusions

not out of reach anymore. Although OmpG samples can already be obtained with a decent degree of order, no proton back-substituted, PRE-doped samples are available so far and the presented spectra only give a first impression of the obtainable spectral quality.

Although the much smaller Aβ^{1-40} was supposed to give rise to a fast assignment and structure calculation, this probe turned out to be a rather improper system for the application of the methodology developed in the thesis. The great advantage of deuteration are potentially extremely narrow resonances that are not broadened by strong dipolar interaction to the proton bath in case of proton dilution. However, due to the intrinsic line broadening of an inhomogeneous sample as expected for a fibrillar (in contrast to a micro-crystalline) protein, the approach can impossibly reach its entire potential. Consequently, for the Aβ^{1-40} peptide, a remaining large resonance overlap and a resulting low sensitivity complicated the usage of the NMR data. Although much work has been dedicated to the sample preparation, all protein samples obtained suffered from large heterogeneity. For this reason, only parts of their spectra overlap, while new resonances appeared and others disappeared with each new preparation. Consequently, unambiguous resonance assignment was hampered and only parts of the primary structure could be assigned so far. Once difficulties on the sample preparation side have been solved, e.g. by an even slower fibrillization process, the assignment using proton detected methodology is supposed to be as simple as for the SH3 domain.

In summary, the presented methodology provides a wide range of new tools for solid state NMR on perdeuterated proteins. Backbone and sidechain assignment experiments are mandatory for most successive spectroscopic work. Although there are still potential extensions, the presented set of experiments thus enables a large improvement of solid state NMR on proteins. This is even more decisive as all mentioned strategies for a resonance assignment and structural constraints are based on an extensive proton dilution and very narrow resonance linewidths. In contrast to solution, the linewidth in the solid state does not depend on the molecular weight of the sample. Especially for large proteins, the potential resolution of solid state NMR can easily rival solution NMR. Thus, the methods will be of utter importance also for membrane proteins whose detergent micelles would tumble too slowly for solution NMR.

The combination of adequate sample preparation with the spectroscopic methodology proposed here gives rise to very much facilitated protein solid state NMR. In addition to the described preliminary applications, these strategies are expected to greatly improve future solid state NMR in general by providing valuable tools for NMR based structure determination and mechanistic understanding of proteins. This way, the margin of accessible targets can be pushed towards larger molecules, those that are difficult to obtain in large amounts, and those that display specific func-

Discussion and conclusions

tion only in a native environment. Although a work like this can always be just a piece of the scientific progress, it will help to better understand many biological processes which are interesting and important from a purely scientific or a medical point of view.

REFERENCES

1. Schanda, P., Huber, M., Verel, R., Ernst, M. & Meier, B. H. Direct Detection of $^{3h}J_{NC'}$ Hydrogen-Bond Scalar Couplings in Proteins by Solid-State NMR Spectroscopy. *Angew. Chem. Int. Ed.* **48**, 9322-9325 (2009).
2. Levinthal, C. Are there pathways for protein folding? *J. Chim. Phys.* **65**, 44-45 (1968).
3. Duan, Y., Wang, L. & Kollman, P. A. The early stage of folding of villin headpiece subdomain observed in a 200-nanosecond fully solvated molecular dynamics simulation. *Proc. Natl. Acad. Sci. USA* **95**, 9897-9902 (1998).
4. Mayhew, M. et al. Protein folding in the central cavity of the GroEL-GroES chaperonin complex. *Nature* **379**, 420-426 (1996).
5. Ferbitz, L. et al. Trigger factor in complex with the ribosome forms a molecular cradle for nascent proteins. *Nature* **131**, 590-596 (2004).
6. Merz, F. et al. Molecular mechanism and structure of trigger factor bound to the translating ribosome. *EMBO J.* **27**, 1622-1632 (2008).
7. Huntington, G. On Chorea. *The Medical and Surgical Reporter: A Weekly Journal* **26**, 317-321 (1872).
8. Alzheimer, A. Über eine eigenartige Erkrankung der Hirnrinde. *Allgemeine Zeitschrift für Psychiatrie und Psychisch-Gerichtliche Medizin* **64**, 146-148 (1907).
9. Prusiner, S. B., Scott, M. R., DeArmond, S. J. & Cohen, F. E. Prion protein biology. *Cell* **93**, 337-348 (1998).
10. Lange, O. F. et al. Recognition Dynamics Up to Microseconds Revealed from an RDC-Derived Ubiquitin Ensemble in Solution. *Science* **320**, 1471-1475 (2008).
11. Bayrhuber, M. et al. Structure of the human voltage-dependent anion channel. *Proc. Natl. Acad. Sci. USA* **105**, 15370–15375 (2008).
12. Hiller, S. et al. Solution Structure of the Integral Human Membrane Protein VDAC-1 in Detergent Micelles. *Science* **321**, 1206 (2008).
13. Cramer, P. et al. Architecture of RNA polymerase II and implications for the transcription mechanism. *Science* **288**, 640-649 (2000).
14. Ruitenberg, M., Kannt, A., Bamberg, E., Fendler, K. & Michel, H. Reduction of cytochrome c oxidase by a second electron leads to proton translocation. *Nature* **417**, 99-102 (2002).
15. Park, J. H., Scheerer, P., Hofmann, K. P., Choe, H.-W. & Ernst, O. P. Crystal structure of the ligand-free G-protein-coupled receptor opsin. *Nature* **454**, 183-187 (2008).
16. Rosenbaum, D. M. et al. GPCR engineering yields high-resolution structural insights into beta2-adrenergic receptor function. *Science* **318**, 1266-1273 (2007).
17. Tugarinov, V., Muhandiram, R., Ayed, A. & Kay, L. E. Four-Dimensional NMR Spectroscopy of a 723-Residue Protein: Chemical Shift Assignments and Secondary Structure of Malate Synthase G. *J. Am. Chem. Soc.* **124**, 10025-10035 (2002).
18. Sprangers, R. & Kay, L. E. Quantitative Dynamics and Binding Studies of the 20S Proteasome by NMR. *Nature* **445**, 618-622 (2008).
19. Kay, L. E., Ikura, M., Tschudin, R. & Bax, A. Three-dimensional triple-resonance NMR spectroscopy of isotopically enriched proteins. *J. Magn. Reson.* **89**, 496-514 (1990).
20. Jeener, J., Meier, B. H., Bachmann, P. & Ernst, R. R. Investigation of exchange processes by two-dimensional NMR spectroscopy. *J. Chem. Phys.* **71** (1979).

References

21. Wishart, D. S., Sykes, B. D. & Richards, F. M. The chemical shift index: a fast and simple method for the assignment of protein secondary structure through NMR spectroscopy. *Biochemistry* **31** (1992).
22. Reif, B., Hennig, M. & Griesinger, C. Direct Measurement of Angles Between Bond Vectors in High-Resolution NMR. *Science* **276**, 1230-1233 (1997).
23. Cordier, F. & Grzesiek, S. Direct Observation of Hydrogen Bonds in Proteins by Interresidue 3hJNC' Scalar Couplings. *J. Am. Chem. Soc.* **121**, 1601-1602 (1999).
24. Tjandra, N., Grzesiek, S. & Bax, A. Magnetic Field Dependence of Nitrogen-Proton J Splittings in ^{15}N-Enriched Human Ubiquitin Resulting from Relaxation Interference and Residual Dipolar Coupling. *J. Am. Chem. Soc.* **118**, 6264 (1996).
25. Henzler-Wildman, K. A. et al. A hierarchy of timescales in protein dynamics is linked to enzyme catalysis. *Nature* **450**, 913-916 (2007).
26. Schanda, P., Forge, V. & Brutscher, B. Protein folding and unfolding studied at atomic resolution by fast two-dimensional NMR spectroscopy. *Proc. Natl. Acad. Sci. USA* **104**, 11257-11262 (2007).
27. Selenko, P., Serber, Z., Gadea, B., Ruderman, J. & Wagner, G. Quantitative NMR analysis of the protein G B1 domain in Xenopus laevis egg extracts and intact oocytes. *Proc. Natl. Acad. Sci. USA* **103**, 11904-11909 (2006).
28. Sakakibara, D. et al. Protein structure determination in living cells by in-cell NMR spectroscopy. *Nature* **458**, 102-105 (2009).
29. Lange, A. et al. Toxin-induced conformational changes in a potassium channel revealed by solid-state NMR. *Nature* **440**, 959-962 (2006).
30. Wasmer, C. et al. Amyloid fibrils of the HET-s(218-289) prion form a beta solenoid with a triangular hydrophobic core. *Science* **319**, 1523-1526 (2008).
31. Loquet, A. et al. 3D Structure Determination of the Crh Protein from Highly Ambiguous Solid-State NMR Restraints. *J. Am. Chem. Soc.* **130**, 3579-3589 (2008).
32. Ernst, R. R., Bodenhausen, G. & Wokaun, A. *Principles of Nuclear Magnetic Resonance in One and Two Dimensions* (Oxford University Press, USA, 1989).
33. Cavanagh, J., Fairbrother, W. J., III, A. G. P., Rance, M. & Skelton, N. J. *Protein NMR Spectroscopy* (Elsevier Academic Press, USA, 2007).
34. Duer, M. *Solid-state NMR spectroscopy: principles and applications* (ed. Duer, M.) (blackwell science, Oxford, 2002).
35. Sorensen, O. W., Eich, G. W., Levitt, M. H., Bodenhausen, G. & Ernst, R. R. Product operator formalism for the description of NMR pulse experiments. *Prog. NMR Spectrosc.* **16**, 163-192 (1983).
36. Delaglio, F., Torchia, D. A. & Bax, A. Measurement of nitrogen-15 carbon-13 J couplings in Staphylococcal nuclease. *J. Biomol. NMR* **1**, 439-446 (1991).
37. Morris, G. A. & Freeman, R. Enhancement of nuclear magnetic resonance signals by polarization transfer. *J. Am. Chem. Soc.* **101**, 760-762 (1979).
38. Tolman, J. R., Flanagan, J. M., Kennedy, M. A. & Prestegard, J. H. Nuclear magnetic dipole interactions in field-oriented proteins: Information for structure determination in solution. *Proc. Natl. Acad. Sci. USA* **92**, 9279–9283 (1995).
39. Tjandra, N. & Bax, A. Direct measurment of distances and angles in biomolecules by nmr in a dilute liquid crystalline medium. *Science* **278**, 1111–1113 (1997).
40. Andrew, E. R., Bradbury, A. & Eades, R. G. Nuclear Magnetic Resonance Spectra from a Crystal rotated at High Speed. *Nature* **182**, 1659 (1958).
41. Baldus, M. Solid-state NMR spectroscopy: molecular structure and organization at the atomic level. *Angew. Chem. Int. Edit.*, 1186-1188 (2006).
42. Raleigh, D. P., Levitt, M. H. & Griffin, R. G. Rotational Resonance in Solid State NMR. *Chem. Phys. Lett.* **146**, 71-76 (1988).

References

43. Bennett, A. E., Ok, J. H., Vega, S. & Griffin, R. G. Chemical shift correlation spectroscopy in rotating solids: Radio frequency-driven dipolar recoupling and longitudinal exchange. *J. Chem. Phys.* **96**, 8624-8627 (1992).
44. Ishii, Y. ^{13}C–^{13}C dipolar recoupling under very fast magic angle spinning in solid-state nuclear magnetic resonance: Applications to distance measurements, spectral assignments, and high-throughput secondary-structure determination. *J. Chem. Phys.* **114**, 8473-8483 (2001).
45. Gullion, T. & Schaefer, J. Rotational-Echo Double-Resonance NMR. *J. Magn. Reson.* **81**, 196-200 (1989).
46. Pines, A., Gibby, M. G. & Waugh, J. S. Proton-enhanced NMR of dilute spins in solids. *J. Chem. Phys.* **59** (1973).
47. Hartmann, S. R. & Hahn, E. L. Nuclear Double Resonance in the Rotating Frame. *Phys. Rev.* **128**, 2042-2053 (1962).
48. Baldus, M., Petkova, A. T., Herzfeld, J. & Griffin, R. G. Cross polarization in the tilted frame: assignment and spectral simplification in heteronuclear spin systems. *Mol. Phys.* **95**, 1197–1207 (1998).
49. Takegoshi, K., Nakamura, S. & Terao, T. ^{13}C–^{1}H dipolar-assisted rotational resonance in magic-angle spinning NMR. *Chem. Phys. Lett.* **344**, 631-637 (2001).
50. Verel, R., Ernst, M. & Meier, B. H. Adiabatic dipolar recoupling in solid-state NMR: the DREAM scheme. *J. Magn. Reson.* **150**, 81-90 (2001).
51. Lewandowski, J. R., DePaepe, G. & Griffin, R. G. Proton Assisted Insensitive Nuclei Cross Polarization. *J. Am. Chem. Soc.* **129**, 728-729 (2007).
52. Levitt, M. H., Kolbert, A. C., Bielecki, A. & Ruben, D. J. High-resolution 1H NMR in solids with frequency-switched multiple-pulse sequences. *Solid state nuclear magnetic resonance* **2**, 151-163 (1993).
53. Vinogradov, E., Madhu, P. K. & Vega, S. Proton spectroscopy in solid state nuclear magnetic resonance with windowed phase modulated Lee-Goldburg decoupling sequences. *Chem. Phys. Lett.* **354**, 193-202 (2002).
54. Brown, S. P., Lesage, A., Elena, B. & Emsley, L. Probing proton-proton proximities in the solid state: High-resolution two-dimensional H-1-H-1 double-quantum CRAMPS NMR spectroscopy. *J. Am. Chem. Soc.* **126**, 13230-13231 (2004).
55. Zhou, D. H. et al. Proton-Detected Solid-State NMR Spectroscopy of Fully Protonated Proteins at 40 kHz Magic-Angle Spinning. *J. Am. Chem. Soc.* **129**, 11791-11801 (2007).
56. McDermott, A. E., Creuzet, F. J., Kolbert, A. C. & Griffin, R. G. High-Resolution Magic-Angle-Spinning NMR Spectra of Protons in Deuterated Solids. *J. Magn. Reson.* **98**, 408-413 (1992).
57. Reif, B. & Griffin, R. G. ^{1}H detected ^{1}H, ^{15}N correlation spectroscopy in rotating solids. *J. Magn. Reson.* **160**, 78-83 (2003).
58. Zhou, D. H. et al. Solid-state protein structure determination with proton-detected triple resonance 3D magic-angle spinning NMR spectroscopy. *Angew. Chem. Int. Ed.* **46**, 8380-8383 (2007).
59. Chevelkov, V., Rehbein, K., Diel, A. & Reif, B. Ultra-High Resolution in Proton Solid-State NMR Spectroscopy at High Levels of Deuteration. *Angew. Chem. Int. Ed.* **45**, 1–5 (2006).
60. Chevelkov, V. et al. Differential Line Broadening in MAS solid-state NMR due to Dynamic Interference. *J. Am. Chem. Soc.* **129**, 10195–10200 (2007).
61. Agarwal, V., Diehl, A. & Skrynnikov, N. High resolution 1H detected 1H,13C correlation spectra in MAS solid-state NMR using deuterated proteins with selective 1H,2H isotopic labeling of methyl groups. *J. Am. Chem. Soc.* **128**, 12620-12621 (2006).

References

62. Agarwal, V. & Reif, B. Residual methyl protonation in perdeuterated proteins for multidimensional correlation experiments in MAS solid-state NMR spectroscopy. *J. Magn. Res.* **194**, 16-24 (2008).
63. Mayer, B. J., Hamaguchi, M. & Hanafusa, H. A novel viral oncogene with structural similarity to phospholipase C. *Nature* **332**, 272-275 (1988).
64. Mayer, B. J. & Hanafusa, H. Mutagenic analysis of the v-crk oncogene: requirement for SH2 and SH3 domains and correlation between increased cellular phosphotyrosine and transformation. *J. Virol.* **64**, 3581-3589 (1990).
65. Bishop, J. Viral oncogenes. *Cell* **42**, 23-28 (1985).
66. Ren, R., Mayer, B. J., Thiel, G. & Baltimore, D. Identification of a 10 amino acid proline-rich SH3 binding sequence. *Science* **259**, 1157-1161 (1993).
67. Kaneko, T., Li, L. & Li, S. S.-C. The SH3 domain- a family of versatile peptide- and protein-recognition module. *Front. Biosci.*, 4938-4952 (2008).
68. Hirai, H. & Varmus, H. E. Site-directed mutagenesis of the SH2- and SH3-coding domains of c-src produces varied phenotypes, including oncogenic activation of p60c-src. *Mol. Cell. Biol.* **10**, 1307-1318 (1990).
69. Jackson, P. & Baltimore, D. N-terminal mutations activate the leukemogenic potential of the myristoylated form of c-abl. *EMBO J.* **8** (1989).
70. Massenet, C. et al. Effects of P47Phox C-Terminus Phosphorylation on Binding Interactions with P40Phox and P67Phox: Structural and Functional Comparison of P40Phox P67Phox SH3 Domains. *J.Biol.Chem.* **280**, 13752 (2005).
71. Castellani, F. et al. Structure of a protein determined by solid-state magicangle-spinning NMR spectroscopy. *Nature* **420**, 98–102 (2002).
72. Chevelkov, V. et al. Detection of dynamic water molecules in a microcrystalline sample of the SH3 domain of α-spectrin by MAS solid-state NMR. *J. Biomol. NMR* **31**, 295–310 (2005).
73. Martin, R. W. & Zilm, K. W. Preparation of protein nanocrystals and their characterization by solid state NMR. *J. Magn. Reson.* **165**, 162–174 (2003).
74. DeLano, W. L. *The PyMOL Molecular Graphics System DeLano Scientific, San Carlos, CA, USA.* (2002).
75. Pauli, J., Baldus, M., van Rossum, B., de Groot, H. & Oschkinat, H. Backbone and Side-Chain 13C and 15N Signal Assignments of the α-Spectrin SH3 Domain by Magic Angle Spinning Solid-State NMR at 17.6 Tesla. *ChemBioChem* **2**, 272-281 (2001).
76. Pauli, J., van Rossum, B.-J., Förster, H., de Groot, H. J. M. & Oschkinat, H. Sample optimization and identification of signal patterns of amino acid side chains in 2D RFDR spectra of the α-spectrin SH3 domain. *J. Magn. Reson.* **143**, 411-416 (2000).
77. Cai, H., Seu, C., Kovacs, Z., Sherry, A. D. & Chen, Y. Sensitivity Enhancement of Multidimensional NMR Experiments by Paramagnetic Relaxation Effects. *J. Am. Chem. Soc.* **128**, 13474-13478 (2006).
78. Bertini, I., Luchinat, C. & Parigi, G. *Solution NMR of Paramagnetic Molecules* (Elsevier, Amsterdam, 2001).
79. Barnhart, J. L. & Berk, R. N. Influence of paramagnetic ions and pH on proton NMR relaxation of biologic fluids. *Invest. Radiol.* **21**, 132–136 (1986).
80. Gillespie, J. R. & Shortle, D. Characterization of long-range structure in the denatured state of staphylococcal nuclease. I. paramagnetic relaxation enhancement by nitroxide spin labels. *J. Mol. Biol.* **268**, 158-169 (1997).
81. Wickramasinghe, N. P., Kotecha, M., Samoson, A., Paast, J. & Y.Ishii. Sensitivity enhancement in 13C solid-state NMR of protein microcrystals by use of paramagnetic metal ions for optimizing 1H T1 relaxation. *J. Magn. Reson.* **184**, 310-316 (2007).

References

82. LaFerla, F. M., Green, K. N. & Oddo, S. Intracellular amyloid-ß in Alzheimer's disease. *Nature Rev.* **8**, 499-509 (2007).
83. Jarrett, J. T., Berger, E. P. & Lansbury, P. T. J. The carboxy terminus of the ß-amyloid protein is critical for the seeding of amyloid formation: Implications for the pathogenesis of Alzheimer's disease. *Biochemistry* **32**, 4693-4697 (1993).
84. C.Haass, Y.Hung, A., G.Schlossmacher, M., Teplow, D. B. & Selkoe, D. J. ß-Amyloid peptide and a 3-kDa fragment are derived by distinct cellular mechanisms. *J. Biol. Chem.* **268**, 3021-3024 (1993).
85. Goedert, M., Wischik, C. M., Crowther, R. A., Walker, J. E. & Klug, A. Cloning and sequencing of the cDNA encoding a core protein of the paired helical filament of Alzheimer disease: identification as the microtubule-associated protein tau. *Proc. Natl. Acad. Sci. USA* **85**, 4051-4055 (1988).
86. Wyss-Coray, T. Inflammation in Alzheimer disease: driving force, bystander or beneficial response? *Nature Med.* **12**, 1005-1015 (2006).
87. Smith, D. G., Cappai, R. & Barnham, K. J. The redox chemistry of the Alzheimer's disease amyloid ß peptide. *Biochim. Biophys. Acta* **1768**, 1976-1990 (2006).
88. Gellermann, G. P. et al. Abeta-globulomers are formed independently of the fibril pathway. *Neurobiol. Dis.* **300**, 212-220 (2008).
89. Petkova, A. T. et al. A structural model for Alzheimer's ß-amyloid fibrils based on experimental constraints from solid state NMR. *Proc. Natl. Acad. Sci. USA* **99**, 16742-16747 (2002).
90. Paravastu, A. K., Leapman, R. D., Yau, W. M. & Tycko, R. Molecular structural Basis for polymorphism in Alzheimer's ß-Amyloid Fibrils. *Proc. Natl. Acad. Sci. USA* **47**, 18349-18354 (2008).
91. Carrio, M., Gonzalez-Montalban, N., Vera, A., Villaverde, A. & Ventura, S. Amyloid-like properties of bacterial inclusion bodies. *J. Mol. Biol.* **347**, 1025-1037 (2005).
92. Wang, L., Maji, S. K., Sawaya, M. R., Eisenberg, D. & Riek, R. Bacterial inclusion bodies contain amyloid-like structure. *PLoS Biol.* **6**, e195 (2008).
93. Teplow, D. B. Preparation of amyloid beta-protein for structural and functional studies. *Methods Enzymol.* **413**, 20-33 (2006).
94. Hou, L., Kang, I., Marchant, R. E. & Zagorski, M. G. Methionine 35 oxidation reduces fibril assembly of the amyloid beta (1-42) peptide of Alzheimers's disease. *J. Biol. Chem.* **277**, 40173-40176 (2002).
95. Wickramasinghe, N. P. et al. Progress in 13C and 1H solid-state Nuclear Magnetic Resonance for Paramagnetic Systems under very fast Magic Angle Spinning. *J. Chem. Phys.* **128**, 052210-052210-15 (2008).
96. Rodriguez-Castañeda, F., Haberz, P., Leonov, A. & Griesinger, C. Paramagnetic tagging of diamagnetic proteins for solution NMR. *Magn. Res. Chem.* **44**, 10-16 (2006).
97. Battiste, J. L. & Wagner, G. Utilization of Site-Directed Spin Labeling and High-Resolution Heteronuclear Nuclear Magnetic Resonance for Global Fold Determination of Large Proteins with Limited Nuclear Overhauser Effect. *Biochemistry* **39**, 5355-5365 (2000).
98. Nadaud, P. S., Helmus, J. J., Kall, S. L. & Jaroniec, C. P. Paramagnetic Ions Enable Tuning of Nuclear Relaxation Rates and Provide Long-Range Structural Restraints in Solid-State NMR of Proteins. *J. Am. Chem. Soc.* **131**, 8108-8120 (2009).
99. Cutting, B., Strauss, A., Fendrich, G., Manley, P. W. & Jahnke, W. NMR resonance assignment of selectively labeled proteins by the use of paramagnetic ligands. *J. Biomol. NMR* **30**, 205–210 (2004).

References

100. Dehner, A. et al. NMR chemical shift perturbation study of the N-terminal domain of Hsp90 upon binding of ADP, AMP-PNP, geldanamycin, and radicicol. *Eur. J. Chem. Biol.* **4**, 870-7 (2003).
101. Schmiedeskamp, M., Rajagopal, P. & Klevit, R. E. NMR chemical shift perturbation mapping of DNA binding by a zinc-finger domain from the yeast transcription factor ADR1. *Protein Sci.* **6**, 1835-1848 (1997).
102. Bertini, I., Luchinat, C. & Parigi, G. Paramagnetic constraints: an aid for quick solution structure determination of paramagnetic metalloproteins. *Concepts in Magnetic Resonance* **14**, 259-286 (2002).
103. Balayssac, S., Bertini, I., Lelli, M., Luchinat, C. & Maletta, M. Paramagnetic Ions Provide Structural Restraints in Solid-State NMR of Proteins. *J. Am. Chem. Soc.* **129**, 2218-2219 (2007).
104. Solomon, I. Relaxation processes in a system of two spins. *Phys. Rev.* **99**, 559-565 (1955).
105. Luginbühl, P. & Wüthrich, K. Semi-classical nuclear spin relaxation theory revisited for use with biological macromolecules. *Prog. NMR Spectrosc.*, 199-247 (2002).
106. Banci, L., Bertini, I. & Luchinat, C. *Nuclear and electron relaxation. The magnetic nucleus-unpaired electron coupling in solution* (VCH, Weinheim, 1991).
107. Bertini, I., Felli, I. C., Luchinat, C., Parigi, G. & Pierattelli, R. Towards a Protocol for Solution Structure Determination of Copper(II) Proteins: the Case of CuIIZnII Superoxide Dismutase. *ChemBioChem* **8**, 1422-1429 (2007).
108. Gueron, M. Nuclear relaxation in macromolecules by paramagnetic ions: a novel mechanism. *J. Magn. Reson.* **19**, 58-66 (1975).
109. Luchinat, C. & Xia, Z. "Paramagnetism and dynamic properties of electrons and nuclei. *Coord. Chem. Rev.* **120**, 281-307 (1992).
110. Pintacuda, G. & Otting, G. Identification of Protein Surfaces by NMR Measurements with a Paramagnetic Gd(III) Chelate. *J. Am. Chem. Soc.* **124**, 457-471 (2002).
111. Linser, R., Chevelkov, V., Diehl, A. & Reif, B. Sensitivity enhancement using paramagnetic relaxation in MAS solid-state NMR of perdeuterated proteins. *J. Magn. Reson.* **189**, 209–216 (2007).
112. Respondek, M., Madl, T., Göbl, C., Golser, R. & Zangger, K. Mapping the Orientation of Helices in Micelle-Bound Peptides by Paramagnetic Relaxation Waves. *J. Am. Chem. Soc.* **129**, 5228-5234 (2007).
113. Petros, A. M., Mueller, L. & Kopple, K. D. NMR Identification of Protein Surfaces using paramagnetic probes. *Biochemistry* **29**, 10045-10048 (1990).
114. Agre, P. et al. Aquaporin CHIP: the archetypal molecular water channel. *Am. J. Physiol. Renal. Physiol.* **265**, F463-F476 (1993).
115. Walz, T. et al. The three-dimensional structure of aquaporin-1. *Nature* **387**, 624-627 (1997).
116. Hetényi, A. et al. Ligand-induced flocculation of neurotoxic fibrillar Abeta(1-42) by noncovalent crosslinking. *ChemBioChem* **9**, 748-757 (2008).
117. Zhu, M., Souillac, P. O., Ionescu-Zanetti, C., Carter, S. A. & Fink, A. L. Surface-catalyzed amyloid fibril formation. *J. Biol. Chem.* **227**, 50914-50922 (2002).
118. Chevelkov, V. & Reif, B. TROSY effects in MAS solid-state NMR. *Concepts NMR Spect.* **32A**, 143–156 (2008).
119. Ayant, Y., Belorizky, E., Fries, P. & Rosset, J. Effet des interactions dipolaires magnétiques intermoléculaires sur la relaxation nucléaire de molécules polyatomiques dans les liquides. *J. Phys.* **38**, 325-337 (1977).
120. Vigouroux, C., Belorizky, E. & Fries, P. H. NMR approach of the electronic properties of the hydrated trivalent rare earth ions in solution. *Eur. Phys. J. D* **5**, 243-255 (1999).

References

121. Connolly, M. L. Analytical Molecular Surface Calculation. *J. Appl. Crystallogr.* **16**, 548 - 558 (1983).
122. Connolly, M. L. Solvent-Accessible Surfaces of Proteins and Nucleic Acids. *Science* **221**, 709 - 713 (1983).
123. Solans, X. & Font-Altaba, M. Crystal Structures of Ethylenediaminetetraacetato Metal Complexes. I. A Comparison of Crystal Structures Containing Hexacoordinated Metal Ions, [(H2O)4X(C10H12 N2O8)Y]n2.nH2O. *Acta Cryst.* **C39**, 435-438 (1983).
124. Bondi, A. Van der Waals volumes and radii. *J. Phys. Chem.* **68**, 441-452 (1964).
125. Otting, G. & Wüthrich, K. Studies of Protein Hydration in Aqueous Solution by Direct NMR Observation of Individual Protein-Bound Water Molecules. *J. Am. Chem. Soc.* **111**, 1871-1875 (1989).
126. Zhou, D. H., Graesser, D. T., Franks, W. T. & Rienstra, C. M. Sensitivity and resolution in proton solid-state NMR at intermediate deuteration levels: Quantitative linewidth characterization and applications to correlation spectroscopy. *J. Magn. Reson.* **178**, 297-307 (2006).
127. Chevelkov, V., Diehl, A. & Reif, B. Measurement of 15N–T1relaxation rates in a perdeuterated protein by MAS solid-stateNMR spectroscopy. *J. Chem. Phys.* **128**, 052316 (2008).
128. Chevelkov, V., Fink, U. & Reif, B. Quantitative analysis of backbone motion in proteins using MAS solid-state NMR spectroscopy. *J. Biomol. NMR* **45**, 197-26 (2009).
129. Akbey, Ü. et al. Optimum levels of exchangeable protons in perdeuterated proteins for proton detection in MAS solid-state NMR spectroscopy. *J. Biomol. NMR* **46**, 67-73 (2009).
130. Siemer, A. B. et al. Observation of Highly Flexible Residues in Amyloid Fibrils of the HET-s Prion. *J. Am. Chem. Soc.* **128**, 13224-13228 (2006).
131. Grzesiek, S. & Bax, A. Improved 3D triple-resonance NMR techniques applied to a 31 kDa protein. *J. Magn. Reson.* **96**, 432–440 (1992).
132. Shaka, A. J., Keeler, J., Frenkiel, T. & Freeman, R. An improved sequence for broadband decoupling: WALTZ-16. *J. Magn. Reson.* **52**, 335-338 (1983).
133. Emsley, L. & Bodenhausen, G. Gaussian pulse cascades: New analytical functions for rectangular selective inversion and in-phase excitation in NMR. *Chem. Phys. Lett.* **165**, 469-476 (1990).
134. Grzesiek, S. & Bax, A. An Efficient Experiment for Sequential Backbone Assignment of Medium-Sized Isotopically Enriched Proteins. *J. Magn. Res.* **99**, 201 (1992).
135. Kay, L. E., Xu, G. Y. & Yamazaki, T. Enhanced-sensitivity triple-resonance spectroscopy with minimal H2O saturation. *J. Magn. Reson.* **A109**, 129–133 (1994).
136. Wittekind, M. & Mueller, L. HNCACB: a high sensitivity 3D NMR experiment to correlate amide proton and nitrogen resonances with the a-carbon and ß-carbon resonances in proteins. *J. Magn. Reson.* **B101** (1993).
137. Yamazaki, T. et al. An HNCA Pulse Scheme for the Backbone Assignment of 15N,13C,2H-Labeled Proteins: application to a 37-kDa Trp Repressor-DNA Complex. **116**, 6464-6465 (1994).
138. Bax, A. & Ikura, M. An efficient 3D NMR technique for correlating the proton and ^{15}N backbone amide resonances with the α-carbon of the preceding residue in uniformly ^{15}N/^{13}C enriched proteins. *J. Biomol. NMR* **1**, 99-104 (1991).
139. Clubb, R. T., Thanabal, V. & Wagner, G. A constant-time threedimensional triple-resonance pulse scheme to correlate intraresidue $^1H^N$, ^{15}N, and ^{13}CO chemical shifts in ^{15}N-^{13}C-labeled proteins. *J. Magn. Reson.* **97**, 213–217 (1992).
140. Engelke, J. & Rüterjans, H. Sequential protein backbone resonance assignments using an improved 3D-HN(CA)CO pulse scheme. *J. Magn. Reson. B* **109**, 318–322 (1995).

References

141. Chevelkov, V., Chen, Z., Bermel, W. & Reif, B. Resolution enhancement in MAS solid-state NMR by application of 13C homonuclear scalar decoupling during acquisition. *J. Magn. Reson.* **172**, 56-62 (2005).
142. Duma, L., Hediger, S., Brutscher, B., Böckmann, A. & Emsley, L. Resolution Enhancement in Multidimensional Solid-State NMR Spectroscopy of Proteins using spin-State Selection. *J. Am. Chem. Soc.* **125**, 11816-11817 (2003).
143. Dayie, K. T. & Wagner, G. Carbonyl Carbon Probe of Local Mobility in $^{13}C,^{15}N$-Enriched Proteins Using High-Resolution Nuclear Magnetic Resonance. *J. Am. Chem. Soc.* **119**, 7797-7806 (1997).
144. Engelke, J. & Rüterjans, H. Backbone dynamics of proteins derived from carbonyl carbon relaxation times at 500, 600 and 800 MHz: Application to ribonuclease T1. *J. Biomol. NMR* **9**, 63-78 (1997).
145. Lipari, G. & Szabo, A. Model-free approach to the interpretation of nuclear magnetic resonance relaxation in macromolecules. 1. Theory and range of validity. *J. Am. Chem. Soc.* **104**, 4546-4559 (1982).
146. Tjandra, N., Feller, S. E., Pastor, R. W. & Bax, A. Rotational diffusion anisotropy of human ubiquitin from 15N NMR relaxation. *J. Am. Chem. Soc.* **117**, 12562-12566 (1995).
147. Renner, C., Baumgartner, R., Noegel, A. A. & Holak, T. A. Backbone Dynamics of the CDK Inhibitor p19INK4d Studied by ^{15}N NMR Relaxation Experiments at Two Field Strengths. *J. Mol. Biol.* **283**, 221-229 (1998).
148. Tugarinov, V., Muhandiram, R., Ayed, A. & Kay, L. E. Four-Dimensional NMR Spectroscopy of a 723-Residue Protein: Chemical Shift Assignments and Secondary Structure of Malate Synthase G. *J. Am. Chem. Soc.* **124**, 10025-10035 (2002).
149. Tugarinov, V., Sprangers, R. & Kay, L. E. Probing Side-Chain Dynamics in the Proteasome by Relaxation Violated Coherence Transfer NMR Spectroscopy. *J. Am. Chem. Soc.* **129**, 1743-1750 (2009).
150. Clore, G. M. et al. Deviations from the Simple 2-parameter model-free approach to the interpretation of N-15 nuclear magnetic relaxation of proteins. *J. Am. Chem. Soc.* **112**, 4989-4991 (1990).
151. Anderson, J. E. & Rawson, D. I. Nuclear magnetic resonance. Slow rotation of a methyl group. *J. Chem. Soc., Chem. Commun.*, 830-831 (1973).
152. Bremi, T., Ernst, M. & Ernst, R. R. Side-Chain Motion with Two Degrees of Freedom in Peptides. An NMR Study of Phenylalanine Side Chains in Antamanide. *J. Phys. Chem.* **98**, 9322-9334 (1994).
153. Markus, M. A., Hinck, A. P., Huang, S., Draper, D. E. & Torchia, D. A. High resolution solution structure of ribosomal protein L22-C76, a helical protein with a flexible loop that becomes structured upon binding to RNA. *Nat. Struct. Biol.* **4**, 70-77 (1997).
154. Chevelkov, V., Fink, U. & Reif, B. Accurate determination of order parameters from 1H,15N dipolar couplings in MAS solid-state NMR experiments. *J. Am. Chem. Soc.* **131**, 14018-14022 (2009).
155. Pervushin, K. V., Riek, R., Wider, G. & Wüthrich, K. Attenuated T2 relaxation by mutual cancellation of dipole-dipole coupling and chemical shift anisotropy indicates an avenue to NMR structures of very large biological macromolecules. *Proc. Natl. Acad. Sci. USA* **94**, 12366-12371 (1997).
156. Yang, D. & Kay, L. E. Improved ^{1}HN-detected triple resonance TROSY-based experiments. *J. Biomol. NMR* **13**, 3-10 (1999).
157. Yang, D. & Kay, L. E. TROSY triple-resonance four-dimensional NMR spectroscopy of a 46 ns tumbling protein. *J. Am. Chem. Soc.* **121**, 2571–2575 (1999).

References

158. Palmer, A. G., Cavanagh, J., Wright, P. E. & Rance, M. Sensitivity Improvement in Proton-Detected Two-Dimensional Heteronuclear Correlation NMR Spectroscopy. *J. Magn. Reson.* **93**, 151-170 (1991).
159. Agarwal, V. & Reif, B. Residual methyl protonation in perdeuterated proteins for multidimensional correlation experiments in MAS solid-state NMR spectroscopy. *J. Magn. Res.* **194**, 16-24 (2008).
160. Leppert, J., Ohlenschlager, O., Gorlach, M. & Ramachandran, R. Adiabatic TOBSY in rotating solids. *J. Biomol. NMR* **29**, 167–173 (2004).
161. Hardy, E. H., Verel, R. & Meier, B. H. Fast MAS total through-bond correlation spectroscopy. *J. Magn. Reson.* **148**, 459–464 (2001).
162. Zhou, D. H. & Rienstra, C. M. High-Performance Solvent Suppression for Proton-Detected Solid-State NMR. *J. Magn. Reson.* **192**, 167-172 (2008).
163. Markley, J. L., Meadows, D. H. & Jardetzk, O. Nuclear Magnetic Resonance Studies of Helix-Coil Transitions in Polyamino Acids. *J. Mol. Biol.* **27**, 25-40 (1967).
164. Wishart, D. S. & Sykes, B. D. Chemical Shifts as a Tool for Structure Determination. *Meth. Enzymol.* **239**, 363-392 (1994).
165. Cornilescu, G., Delaglio, F. & Bax, A. Protein backbone angle restraints from searching a database for chemical shift and sequence homology. *J. Biomol. NMR* **13**, 289-302 (1999).
166. Manolikas, T., Herrmann, T. & Meier, B. H. Protein Structure Determination from 13C Spin-Diffusion Solid-State NMR Spectroscopy. *J. Am. Chem. Soc.* **130**, 3959 – 3966 (2008).
167. Lange, A. et al. A concept for rapid protein-structure determination by solid-state NMR spectroscopy. *Angew. Chem. Int. Ed.* **44**, 2089-2092 (2006).
168. Grzesiek, S., Anglister, J., Ren, H. & Bax, A. ^{13}C line narrowing by deuterium decoupling in ^{2}D/^{13}C/^{15}N enriched proteins. Application to triple resonance 4D J connectivity of sequential amides. *J. Am. Chem. Soc.* **115**, 4369-4370 (1993).
169. Sun, Z.-Y., Frueh, D., Selenko, P., Hoch, J. & Wagner, G. Fast Assignment of 15N-HSQC Peaks using High-Resolution 3D HNcocaNH Experiments with Non-Uniform Sampling. *J. Biomol. NMR* **33**, 43-50 (2005).
170. Ikura, M., Bax, A., Clore, G. M. & Gronenborn, A. M. Detection of nuclear Overhauser effects between degenerate amide proton resonances by heteronuclear three-dimensional NMR spectroscopy. *J. Am. Chem. Soc.* **112**, 9020-9022 (1990).
171. Kay, L. E., Clore, G. M., Bax, A. & Gronenborn, A. M. Four-dimensional heteronuclear triple-resonance NMR spectroscopy of interleukin-1 beta in solution. *Science* **249** (1990).
172. Bracken, C., Palmer III, A. G. & Cavanagh, J. (H)N(COCA)NH and HN(COCA)NH experiments for 1H-15N backbone assignments in 13C/15N-labeled proteins. *J. Biomol. NMR* **9**, 94–100 (1997).
173. Rossum, B. v. et al. Assignment of amide proton signals by combined evaluation of HN, NN and HNCA MAS-NMR correlation spectra. *J. Biomol. NMR* **25**, 217–223 (2003).
174. Krushelnitsky, A. et al. Direct Observation of Millisecond to Second Motions in Proteins by Dipolar CODEX NMR Spectroscopy. *J. Am. Chem. Soc.* **131**, 12097-12099 (2009).
175. Reif, B. et al. Protein side-chain dynamics observed by solution- and solid-state NMR: comparative analysis of methyl 2H relaxation data. *J. Am. Chem. Soc.* **128**, 12354-12355 (2006).
176. Rossum, B.-J. v., Castellani, F., Rehbein, K., Pauli, J. & Oschkinat, H. Assignment of the Nonexchanging Protons of the α-Spectrin SH3 Domain by Two- and Three-Dimensional ^{1}H-^{13}C Solid-State Magic-Angle Spinning NMR and Comparison of Solution and Solid-State Proton Chemical Shifts. *ChemBioChem* **2**, 906-914 (2001).
177. Castellani, F. (Thesis, Freie Universität, Berlin, 2003).

References

178. Salikhov, K. M., Molin, Y. N., Sagdeev, Z. R. & Buchachenko, A. L. *Spin Polarization and Magnetic Effects in Radical Reactions* (ed. Molin, Y. N.) (Elsevier, Amsterdam, 1984).
179. Hensley, K. et al. A model for ß-amyloid aggregation and neurotoxicity based on free radical generation by the peptide: Relevance to Alzheimer disease. *Proc. Natl. Acad. Sci. U.S.A.* **91**, 3270-3274 (1993).

APPENDIX

Out-and-back HNCO using INEPTs and WALTZ-16 decoupling

```
;rl_hncosel3test
;3D sequence with
;   inverse correlation for triple resonance using multiple
;      inept transfer steps
;
;      F1(H) -> F3(N) -> F2(C=O,t1) -> F3(N,t2) -> F1(H,t3)
;
;on/off resonance Ca and C=O pulses using soft/shaped pulses
;phase sensitive (t1)
;phase sensitive (t2)
;using constant time in t2


prosol relations=<triple>


#include <Avance.incl>
#include <Delay.incl>



"d0=3u"

"d11=30m"

"d13=4u"
"d16=4u"

"d21=5.5m"
"d23=12m"
"d26=2.3m"


"in29=in10"
"in30=in10"

"d10=d23/2-p14/2"
"d29=d23/2-p14/2-p26-d21-4u"
"d30=d23/2-p14/2"


"DELTA=d0*2+larger(p14,p22)-p14"
"DELTA1=d23-d21-p26"
"DELTA2=d26-p16-d16-p11-12u"



aqseq 321


1 d11 ze
  d11 pl16:f3
2 d11 do:f3
```

Appendix

```
3 d1 pl1:f1
  2u pl2:f2
  p1 ph1
  d26 pl3:f3
  (center (p2 ph1) (p22 ph1):f3 )
  d26
  (p1 ph2):f1

  4u pl0:f1
  4u
  (p21 ph3):f3
  d21
  4u pl19:f1
  DELTA1 cpds1:f1 ph1
  4u
  (center (p4 ph1):f2 (p22 ph1):f3 )
  d23  pl4:f2
  (p21 ph1):f3
  10u do:f1
  (p3 ph4):f2
  d0
  (center (p14:sp5 ph1):f2 (p22 ph1):f3 )
  d0
  4u   pl2:f2
  (p4 ph1):f2
  DELTA
  (p14:sp5 ph1):f2
  4u
  4u   pl4:f2
  (p3 pl4 ph1):f2
  4u
  10u cpds1:f1 ph1
  (p21 ph5):f3
  d30
  (p14:sp5 ph1):f2
  d30
  4u pl2:f2
  (center (p4 ph1):f2   (p22 ph8):f3)
  d10
  (p14:sp5 ph1):f2
  d29
  4u do:f1
  4u
  d21
  (p21 ph1):f3

  d16
  4u
  4u pl1:f1

  (p1 ph1)
  4u
  d26
  4u
  4u pl1:f1
  (center (p2 ph1) (p22 ph1):f3 )
  4u

  4u
  d26            pl16:f3
```

Appendix

```
  4u
  go=2 ph31 cpd3:f3
  d11 do:f3 mc #0 to 2
     F1PH(rd10 & rd29 & rd30 & ip4, id0)
     F2PH(ip5, id10 & id29 & dd30)
exit


ph1=0
ph2=1
ph3=0 0 0 0 0 0 0 0 2 2 2 2 2 2 2 2
ph4=0 2
ph5=0 0 2 2
ph6=2
ph7=3
ph8=0 0 0 0 2 2 2 2
ph31=0 2 2 0 0 2 2 0 2 0 0 2 2 0 0 2


;pl0 : 120dB
;pl1 : f1 channel - power level for pulse (default)
;pl2 : f2 channel - 180 soft rectang. C
;pl3 : f3 channel - power level for pulse (default)
;pl4 : f2 channel - 90 soft rectang. C
;pl16: f3 channel - power level for CPD/BB decoupling
;pl19: f1 channel - power level for CPD/BB decoupling
;sp5: f2 channel - shaped pulse 180 degree  (Ca off resonance)
;p1 : f1 channel -  90 degree high power pulse
;p2 : f1 channel - 180 degree high power pulse
;p3: f2 channel -  90 degree soft pulse
;p14: f2 channel - 180 degree shaped pulse
;p4: f2 channel - 180 degree soft pulse
;p21: f3 channel -  90 degree high power pulse
;p22: f3 channel - 180 degree high power pulse
;d0 : incremented delay (F1 in 3D)              [3 usec]
;d1 : relaxation delay; 1-5 * T1
;d10: incremented delay (F2 in 3D) =  d23/2-p14/2
;d11: delay for disk I/O                        [30 msec]
;d13: short delay                               [4 usec]
;d16: short delay
;d21: 1/(2J(NH)                                 [5.5 msec]
;d23: 1/(4J(NCO)                                [12 msec]
;d26: 1/(4J'(NH)                                [2.3 msec]
;d29: incremented delay (F2 in 3D) = d23/2-p14/2-p26-d21-4u
;d30: decremented delay (F2 in 3D) = d23/2-p14/2
;cnst21: CO chemical shift (offset, in ppm)
;cnst22: Calpha chemical shift (offset, in ppm)
;o2p: CO chemical shift (cnst21)
;in0: 1/(2 * SW(CO)) =  DW(CO)
;nd0: 2
;in10: 1/(4 * SW(N)) = (1/2) DW(N)
;nd10: 4
;in29: = in10
;in30: = in10
;NS: 8 * n
;td1: number of experiments in F1
;td2: number of experiments in F2       td2 max = 2 * d30 / in30
;FnMODE: States-TPPI (or TPPI) in F1
;FnMODE: States-TPPI (or TPPI) in F2
;cpds1: decoupling according to sequence defined by cpdprg1
```

Appendix

```
;cpd3: decoupling according to sequence defined by cpdprg3
;pcpd1: f1 channel - 90 degree pulse for decoupling sequence
;pcpd3: f3 channel - 90 degree pulse for decoupling sequence
```

Out-and-back HNCO using CPs and WALTZ-16 decoupling

```
;rl_hncoCP2test_water
;3D sequence with
;    inverse correlation for triple resonance using multiple
;       inept transfer steps
;
;     F1(H) -> F3(N) -> F2(C=O,t1) -> F3(N,t2) -> F1(H,t3)
;
;on/off resonance Ca and C=O pulses using soft/shaped pulses
;phase sensitive (t1)
;phase sensitive (t2)
;using constant time in t2
;water suppression using mississippi pulses (Zhou and Rienstra, 2008)

;$OWNER=guest
prosol relations=<triple>


#include <Avance.incl>
#include <Delay.incl>


"d0=3u"

"d11=30m"

"d13=4u"

"d21=5.5m"
"d23=12m"
"d26=2.3m"


"in29=in10"
"in30=in10"

"d10=d23/2-p14/2"
"d29=d23/2-p14/2-p26-d21-4u"
"d30=d23/2-p14/2"


"DELTA=d0*2+larger(p14,p22)-p14"
"DELTA1=d23-d21-p26-2u"
"DELTA2=d26-p16-d16-p11-12u"


aqseq 321
```

Appendix

```
1 d11 ze
  d11 pl16:f3
2 d11 do:f3
3 d1 pl1:f1
20u reset:f1 reset:f3  reset:f2
3u fq=cnst3:f1
  2u pl2:f2
  2u pl17:f3
  p1 ph1
  0.4u pl5:f1

  (p15:spf1 ph3):f3 (p15 ph2):f1

  4u
  4u pl19:f1
  4u

  4u   cpds1:f1 ph2

  d21
  2u pl3:f3

  DELTA1
  4u
  (center (p4 ph1):f2 (p22 ph1):f3 )
  d23  pl4:f2

  (p21 ph2):f3

  10u do:f1

  (p3 ph4):f2
  d0
  (center (p14:sp5 ph2):f2 (p22 ph1):f3 )
  d0
  4u   pl2:f2
  (p4 ph1):f2
  DELTA
  (p14:sp5 ph1):f2
  4u
  4u   pl4:f2
  (p3 pl4 ph1):f2

  4u pl19:f1
  4u cpds1:f1 ph2

  (p21 ph5):f3
  d30

  (p14:sp5 ph1):f2

  d30
  4u pl2:f2

  (center (p4 ph1):f2  (p22 ph8):f3)
  d10

  (p14:sp5 ph1):f2

  d29
```

Appendix

```
  d21

  (p21 ph1):f3

  10u    do:f1
4u pl10:f1
(p10 ph2):f1
(p10 ph1):f1
(p10 ph2):f1
(p10 ph1):f1
4u
(p21 ph6):f3

  4u pl17:f3
  4u pl5:f1

  (p16:spf2 ph2):f3   (p16 ph1):f1
  2u       pl16:f3

  4u
  go=2 ph31 cpd3:f3
  d11 do:f3 mc #0 to 2
     F1PH(rd10 & rd29 & rd30 & ip4, id0)
     F2PH(ip5, id10 & id29 & dd30)
exit


ph1=0
ph2=1
ph3=1 1 1 1 1 1 1 3 3 3 3 3 3 3
ph4=0 2
ph5=1 1 3 3
ph6=2
ph7=3
ph8=0 0 0 0 2 2 2 2
ph31=0 2 2 0 0 2 2 0 2 0 0 2 2 0 0 2


;p10 : 120dB
;pl1 : f1 channel - power level for pulse (default)
;pl2 : f2 channel - 180 soft rectang. C
;pl3 : f3 channel - power level for pulse (default)
;pl4 : f2 channel - 90 soft rectang. C
;pl10: f1 channel - power level water suppression (10kHz)
;pl16: f3 channel - power level for CPD/BB decoupling
;pl19: f1 channel - power level for CPD/BB decoupling
;pl5:  f1 channel - power level for CP
;pl17: f3 channel - power level for CP
;sp5: f2 channel - shaped pulse 180 degree  (Ca off resonance)
;p1 : f1 channel -  90 degree high power pulse
;p2 : f1 channel - 180 degree high power pulse
;p10: water purge pulses (50ms each)
;p3: f2 channel -  90 degree soft pulse
;p4: f2 channel - 180 degree soft pulse
;p14: f2 channel - 180 degree shaped pulse
;p15: first HN CP                [1.5 msec]
;p16: second HN CP               [1.5 msec]
;p21: f3 channel -  90 degree high power pulse
;p22: f3 channel - 180 degree high power pulse
```

Appendix

```
;d0 : incremented delay (F1 in 3D)                       [3 usec]
;d1 : relaxation delay; 1-5 * T1
;d10: incremented delay (F2 in 3D) = d23/2-p14/2
;d11: delay for disk I/O                                 [30 msec]
;d13: short delay                                        [4 usec]
;d21: 1/(2J(NH))                                         [5.5 msec]
;d23: 1/(4J(NCO))                                        [12 msec]
;d26: 1/(4J'(NH))                                        [2.3 msec]
;d29: incremented delay (F2 in 3D) = d23/2-p14/2-p26-d21-4u
;d30: decremented delay (F2 in 3D) = d23/2-p14/2
;cnst21: CO chemical shift (offset, in ppm)
;cnst22: Calpha chemical shift (offset, in ppm)
;o2p: CO chemical shift (cnst21)
;in0: 1/(2 * SW(CO)) = DW(CO)
;nd0: 2
;in10: 1/(4 * SW(N)) = (1/2) DW(N)
;nd10: 4
;in29: = in10
;in30: = in10
;NS: 8 * n
;DS: >= 16
;td1: number of experiments in F1
;td2: number of experiments in F2       td2 max = 2 * d30 / in30
;FnMODE: States-TPPI (or TPPI) in F1
;FnMODE: States-TPPI (or TPPI) in F2
;cpds1: decoupling according to sequence defined by cpdprg1
;cpd3: decoupling according to sequence defined by cpdprg3
;pcpd1: f1 channel - 90 degree pulse for decoupling sequence
;pcpd3: f3 channel - 90 degree pulse for decoupling sequence
```

Out-and-back TROSY -HNCO

```
;r1_hncoTrosy_purge
;3D sequence with
;    inverse correlation for triple resonance using multiple
;       inept transfer steps
;
;       F1(H) -> F3(N) -> F2(C=O,t1) -> F3(N,t2) -> F1(H,t3)
;
;on/off resonance Ca and C=O pulses using soft/shaped pulses
;phase sensitive (t1)
;phase sensitive (t2)
;using constant time in t2
;water suppression using purge pulse in first HN-INEPT


prosol relations=<triple>


#include <Avance.incl>
```

Appendix

```
#include <Delay.incl>



"d0=3u"

"d10=3u"

"d11=30m"

"d12=20u"


"d23=12m"
"d26=2.3m"


"in30=in10"

"d30=d23-p1"


"DELTA=d0*2+larger(p14,p22)-p14"
"DELTA1=d26-p11-p16-d16-8u"
"DELTA2=d23-d10-p14-p21*4/3.1416-4u"
"DELTA3=d26-p16-d16"



"l0=1"

"spoff2=0"
"spoff3=0"
;"spoff5=bf2*(cnst22/1000000)-o2"
"spoff8=0"


aqseq 321


1 d11 ze
  d11 pl16:f3
2 d11 do:f3
3 d1 pl1:f1
  2u pl2:f2
  p1 ph1
  d26 pl3:f3
  (center (p2 ph2) (p22 ph1):f3 )
  d26
  (p5 ph1):f1
  1u
  (p1 ph4):f1

  4u
  (p21 ph1):f3
  d23
  (center (p4 ph1):f2 (p22 ph1):f3 )
  d23   pl4:f2
  (p21 ph2):f3
```

Appendix

```
  (p3 ph5):f2
  d0
  (center (p14:sp5 ph1):f2 (p22 ph1):f3 )
  d0
  4u   pl2:f2
  (p4 ph1):f2
  DELTA
  (p14:sp5 ph1):f2
  4u
  4u   pl4:f2
  (p3 p14 ph1):f2

  4u
   if "l0 %2 == 1"
      {
      (p21 ph6):f3
      }
   else
      {
      (p21 ph7):f3
      }

  d10
  (p14:sp5 ph1):f2
  DELTA2
  4u pl2:f2
  (center (p4 ph1):f2   (p22 ph1):f3)
  d30
  (p1 ph8)

  d26

  (center (p2 ph1) (p22 ph1):f3 )

  d26
  (center (p1 ph1) (p21 ph8):f3 )
  d26
  (center (p2 ph1) (p22 ph1):f3 )
  d26
  (p21 ph1):f3
  go=2 ph31
  d11 do:f3 mc #0 to 2
      F1PH(rd10 & rd30 & ip5, id0)
      F2EA(ip8*2 & iu0, id10 & dd30)
exit


ph1=0
ph2=1
ph3=2
ph4=3
ph5=0 0 0 0 2 2 2
ph6=1 3 0 2
ph7=1 3 2 0
ph8=3
ph31=1 3 2 0 3 1 0 2


;pl0 : 120dB
;pl1 : f1 channel - power level for pulse (default)
;pl2 : f2 channel - 180 soft rectang. C
```

Appendix

```
;p13 : f3 channel - power level for pulse (default)
;p14 : f2 channel - 90 soft rectang. C
;pl16: f3 channel - power level for CPD/BB decoupling
;pl19: f1 channel - power level for CPD/BB decoupling
;sp5: f2 channel - shaped pulse 180 degree  (Ca off resonance)
;p1 : f1 channel -  90 degree high power pulse
;p2 : f1 channel - 180 degree high power pulse
;p5: f1 channel -  water purge pulse           [1 msec]
;p13: f2 channel -  90 degree shaped pulse
;p14: f2 channel - 180 degree shaped pulse
;p21: f3 channel -  90 degree high power pulse
;p22: f3 channel - 180 degree high power pulse
;d0 : incremented delay (F1 in 3D)              [3 usec]
;d1 : relaxation delay; 1-5 * T1
;d10: incremented delay (F2 in 3D) =  d23/2-p14/2
;d11: delay for disk I/O                        [30 msec]
;d13: short delay                               [4 usec]
;d21: 1/(2J(NH)                                 [5.5 msec]
;d23: 1/(4J(NCO)                                [12 msec]
;d26: 1/(4J'(NH)                                [2.3 msec]
;d29: incremented delay (F2 in 3D) = d23/2-p14/2-p26-d21-4u
;d30: decremented delay (F2 in 3D) = d23/2-p14/2
;cnst21: CO chemical shift (offset, in ppm)
;cnst22: Calpha chemical shift (offset, in ppm)
;o2p: CO chemical shift (cnst21)
;in0: 1/(2 * SW(CO)) =  DW(CO)
;nd0: 2
;in10: 1/(2 * SW(N)) =  DW(N)
;nd10: 2
;in29: = in10
;in30: = in10
;NS: 8 * n
;DS: >= 16
;td1: number of experiments in F1
;td2: number of experiments in F2    td2 max = 2 * d30 / in30
;FnMODE: States-TPPI (or TPPI) in F1
;FnMODE: echo/antiecho in F2
;cpds1: decoupling according to sequence defined by cpdprg1
;cpd3: decoupling according to sequence defined by cpdprg3
;pcpd1: f1 channel - 90 degree pulse for decoupling sequence
;pcpd3: f3 channel - 90 degree pulse for decoupling sequence
```

Out-and-back HNCACB using INEPTs and WALTZ-16 decoupling

```
;rl_hncacbtest
;HNCACB
;3D sequence with
;    inverse correlation for triple resonance using multiple
;       inept transfer steps
;
;    F1(H) -> F3(N) -> F2(Ca -> Cb,t1) -> F3(N,t2) -> F1(H,t3)
;
;on/off resonance Ca and C=O pulses using soft/shaped pulses
```

Appendix

```
;phase sensitive (t1)
;phase sensitive (t2)
;using constant time in t2

prosol relations=<triple>


#include <Avance.incl>
#include <Delay.incl>



"d0=3u"

"d11=30m"

"d13=4u"
"d16=4u"

"d21=5.5m"
"d23=12.4m"
"d26=1.95m"
"d28=3.6m"


"in29=in10"
"in30=in10"

"d10=d23/2-p14/2"
"d29=d23/2-p14/2-p26-d21-4u"
"d30=d23/2-p14/2"


"DELTA1=d23-d21-p26"
"DELTA2=d0*2+larger(p14,p22)-p14"


aqseq 321

1 ze
  d11 pl16:f3
2 d11 do:f3
3 d1 pl1:f1
  2u pl2:f2
  (p1 ph1)
  d26 pl3:f3
  (center (p2 ph1) (p22 ph1):f3 )
  d26
  (p1 ph2):f1

  4u pl0:f1

  4u

  (p21 ph3):f3
  d21 pl19:f1
  DELTA1 cpds1:f1 ph1
  4u
  (center (p4 ph1):f2 (p22 ph1):f3 )
  d23    pl4:f2
```

Appendix

```
  (p21 ph1):f3

4u do:f1

  (p3 ph4):f2
  d28 pl2:f2
  (p4 ph1):f2
  d28  pl4:f2
  (p3 ph2):f2

  d0
  (center (p14:sp5 ph1):f2 (p22 ph8):f3 )
  d0
  4u pl2:f2
  (p4 ph1):f2
  DELTA2
  (p14:sp5 ph1):f2
  4u
  4u pl4:f2
  (p3 ph9):f2
  d28 pl2:f2
  (p4 ph1):f2
  d28 pl4:f2
  (p3 ph10):f2


  4u

    4u pl19:f1
    4u cpds1:f1 ph1

  (p21 ph5):f3
  d30
  (p14:sp5 ph1):f2
  d30
  4u pl2:f2
  (center (p4 ph1):f2 (p22 ph8):f3 )
  d10
  (p14:sp5 ph1):f2
  d29
  4u do:f1
  d21
  (p21 ph1):f3


  d16
  4u pl1:f1

  (p1 ph1)

  d26   pl1:f1
  (center (p2 ph1) (p22 ph1):f3 )
  d26        pl16:f3

  4u
  go=2 ph31 cpd3:f3
  d11 do:f3 mc #0 to 2
     F1PH(rd10 & rd29 & rd30 & ip9 & ip10, id0 & dp9*2)
     F2PH(ip5, id10 & id29 & dd30)
exit
```

Appendix

```
ph1=0
ph2=1
ph3=0 0 0 0 0 0 0 0 2 2 2 2 2 2 2 2
ph4=0
ph5=0 0 2 2
ph6=2
ph7=3
ph8=0 0 0 0 2 2 2 2
ph9=3 1
ph10=0 2
ph31=0 2 2 0 0 2 2 0 2 0 0 2 2 0 0 2


;pl0 : 120dB
;pl1 : f1 channel - power level for pulse (default)
;pl3 : f3 channel - power level for pulse (default)
;pl16: f3 channel - power level for CPD/BB decoupling
;pl19: f1 channel - power level for CPD/BB decoupling
;sp5: f2 channel - shaped pulse 180 degree   (C=O off resonance)
;p1 : f1 channel -  90 degree high power pulse
;p2 : f1 channel - 180 degree high power pulse
;p3: f2 channel -  90 degree soft pulse
;p4: f2 channel - 180 degree soft pulse
;p14: f2 channel - 180 degree shaped pulse
;p21: f3 channel -  90 degree high power pulse
;p22: f3 channel - 180 degree high power pulse
;d0 : incremented delay (F1 in 3D)             [3 usec]
;d1 : relaxation delay; 1-5 * T1
;d10: incremented delay (F2 in 3D) =  d23/2-p14/2
;d11: delay for disk I/O                       [30 msec]
;d13: short delay                              [4 usec]
;d16: short delay
;d21: 1/(2J(NH)                                [5.5 msec]
;d23: 1/(4J(NCa)                               [12.4 msec]
;d26: 1/(4J'(NH)                               [2.3 msec]
;d28: 1/(4J(CaCb)                              [3.6 msec]
;d29: incremented delay (F2 in 3D) = d23/2-p14/2-p26-d21-4u
;d30: decremented delay (F2 in 3D) = d23/2-p14/2
;cnst21: CO chemical shift (offset, in ppm)
;cnst23: Caliphatic chemical shift (offset, in ppm)
;o2p: Caliphatic chemical shift (cnst23)
;in0: 1/(2 * SW(Cali)) =  DW(Cali)
;nd0: 2
;in10: 1/(4 * SW(N)) = (1/2) DW(N)
;nd10: 4
;in29: = in10
;in30: = in10
;NS: 8 * n
;DS: >= 16
;td1: number of experiments in F1
;td2: number of experiments in F2     td2 max = 2 * d30 / in30
;FnMODE: States-TPPI (or TPPI) in F1
;FnMODE: States-TPPI (or TPPI) in F2
;cpds1: decoupling according to sequence defined by cpdprg1
;cpd3: decoupling according to sequence defined by cpdprg3
;pcpd1: f1 channel - 90 degree pulse for decoupling sequence
;pcpd3: f3 channel - 90 degree pulse for decoupling sequence
```

Appendix

Out-and-back HNCACB using CPs and WALTZ-16 decoupling

```
;rl_hncacbCP2test_water
;HNCACB
;3D sequence with
;    inverse correlation for triple resonance using multiple
;       inept transfer steps
;
;       F1(H) -> F3(N) -> F2(C=O,t1) -> F3(N,t2) -> F1(H,t3)
;
;on/off resonance Ca and C=O pulses using soft/shaped pulses
;phase sensitive (t1)
;phase sensitive (t2)
;using constant time in t2
;water suppression using purge pulses à la Zhou and Rienstra 2008

;$OWNER=guest
prosol relations=<triple>


#include <Avance.incl>
#include <Delay.incl>


"d0=3u"

"d11=30m"

"d13=4u"

"d21=5.5m"
"d23=12m"
"d26=2.3m"
"d28=3.6m"


"in29=in10"
"in30=in10"

"d10=d23/2-p14/2"
"d29=d23/2-p14/2-p26-d21-4u"
"d30=d23/2-p14/2"


"DELTA=d0*2+larger(p14,p22)-p14"
"DELTA1=d23-d21-p26-2u"
"DELTA2=d26-p16-d16-p11-12u"


aqseq 321
```

Appendix

```
1 d11 ze
  d11 pl16:f3
2 d11 do:f3
3 d1 pl1:f1
20u reset:f1 reset:f3  reset:f2
3u fq=cnst0:f1
  2u pl2:f2
  2u pl17:f3
  p1 ph1
 0.4u pl15:f1

  (p15:spf1 ph3):f3 (p15 ph2):f1

  3u fq=cnst1:f1
  4u
  4u pl19:f1
  4u

   4u   cpds1:f1 ph2

   d21
   2u pl3:f3

   DELTA1
   4u
   (center (p4 ph1):f2 (p22 ph1):f3 )
   d23   pl4:f2

(p21 ph2):f3

   10u do:f1

   (p3 ph4):f2
   d28 pl2:f2
   (p4 ph1):f2
   d28  pl4:f2
   (p3 ph2):f2

   d0
   (center (p14:sp5 ph1):f2 (p22 ph8):f3 )
   d0
   4u pl2:f2
   (p4 ph1):f2
   DELTA
   (p14:sp5 ph1):f2
   4u
   4u pl4:f2
   (p3 ph9):f2
   d28 pl2:f2
   (p4 ph1):f2
   d28 pl4:f2
   (p3 ph10):f2

   4u pl19:f1
   4u cpds1:f1 ph2

   (p21 ph5):f3
   d30

   (p14:sp5 ph1):f2
```

Appendix

```
  d30
  4u pl2:f2

  (center (p4 ph1):f2   (p22 ph8):f3)
  d10

  (p14:sp5 ph1):f2

  d29

  d21

(p21 ph1):f3

   10u     do:f1
   2u  fq=cnst0:f1
  4u pl10:f1
  (p10 ph2):f1
  (p10 ph1):f1
  (p10 ph2):f1
  (p10 ph1):f1
  4u
  (p21 ph6):f3

  4u pl17:f3
  4u pl5:f1

  (p16:spf2 ph2):f3   (p16 ph1):f1
   2u         pl16:f3

   4u
  go=2 ph31 cpd3:f3
  d11 do:f3 mc #0 to 2
      F1PH(rd10 & rd29 & rd30 & ip9 & ip10, id0 & dp9*2)
      F2PH(ip5, id10 & id29 & dd30)
exit


ph1=0
ph2=1
ph3=1 1 1 1 1 1 1 3 3 3 3 3 3 3
ph4=0
ph5=1 1 3 3
ph6=2
ph7=3
ph8=0 0 0 0 2 2 2 2
ph9=3 1
ph10=0 2
ph31=0 2 2 0 0 2 2 0 2 0 0 2 2 0 0 2


;pl0 : 120dB
;pl1 : f1 channel - power level for pulse (default)
;pl2 : f2 channel - 180 soft rectang. C
;pl3 : f3 channel - power level for pulse (default)
;pl4 : f2 channel - 90 soft rectang. C
;pl10: f1 channel - power level water suppression (10kHz)
;pl16: f3 channel - power level for CPD/BB decoupling
;pl19: f1 channel - power level for CPD/BB decoupling
;pl5: f1 channel - power level for CP
```

Appendix

```
;pl17: f3 channel - power level for CP
;sp5: f2 channel - shaped pulse 180 degree   (Ca off resonance)
;p1 : f1 channel -  90 degree high power pulse
;p2 : f1 channel - 180 degree high power pulse
;p10: water purge pulses (50ms each)
;p3: f2 channel -  90 degree soft pulse
;p4: f2 channel - 180 degree soft pulse
;p14: f2 channel - 180 degree shaped pulse
;p15: first HN CP                           [1.5 msec]
;p16: second HN CP                          [1.5 msec]
;p21: f3 channel -  90 degree high power pulse
;p22: f3 channel - 180 degree high power pulse
;d0 : incremented delay (F1 in 3D)                    [3 usec]
;d1 : relaxation delay; 1-5 * T1
;d10: incremented delay (F2 in 3D) =  d23/2-p14/2
;d11: delay for disk I/O                              [30 msec]
;d13: short delay                                     [4 usec]
;d21: 1/(2J(NH)                                       [5.5 msec]
;d23: 1/(4J(NCO)                                      [12 msec]
;d26: 1/(4J'(NH)                                      [2.3 msec]
;d29: incremented delay (F2 in 3D) = d23/2-p14/2-p26-d21-4u
;d30: decremented delay (F2 in 3D) = d23/2-p14/2
;cnst21: CO chemical shift (offset, in ppm)
;cnst22: Calpha chemical shift (offset, in ppm)
;o2p: CO chemical shift (cnst21)
;in0: 1/(2 * SW(CO)) = DW(CO)
;nd0: 2
;in10: 1/(4 * SW(N)) = (1/2) DW(N)
;nd10: 4
;in29: = in10
;in30: = in10
;NS: 8 * n
;DS: >= 16
;td1: number of experiments in F1
;td2: number of experiments in F2     td2 max = 2 * d30 / in30
;FnMODE: States-TPPI (or TPPI) in F1
;FnMODE: States-TPPI (or TPPI) in F2
;cpds1: decoupling according to sequence defined by cpdprg1
;cpd3: decoupling according to sequence defined by cpdprg3
;pcpd1: f1 channel - 90 degree pulse for decoupling sequence
;pcpd3: f3 channel - 90 degree pulse for decoupling sequence
```

Out-and-back HNCACO/HNCOCA using INEPTs

```
;hncoca
;HN(CA)CO or HN(CO)CA
;3D sequence with
;    inverse correlation for triple resonance using multiple
;       inept transfer steps
;
;    F1(H) -> F3(N) -> F2(C=O) -> F2(Ca,t1)
;           -> F2(C=O) -> F3(N,t2) -> F1(H,t3)
;
;on/off resonance Ca and C=O pulses using soft/shaped pulses
```

Appendix

```
;phase sensitive (t1)
;phase sensitive (t2)
;using constant time in t2
;water suppression using water purge pulse in first INEPT
;$CLASS=HighRes
;$DIM=3D
;$TYPE=
;$SUBTYPE=
;$COMMENT=


prosol relations=<triple>


#include <Avance.incl>
#include <Delay.incl>


"p2=p1*2"
"p22=p21*2"
"d0=3u"

"d11=30m"

"d13=4u"

"d21=5.5m"
"d22=4m"
"d23=12m"
"d26=2.3m"


"in29=in10"
"in30=in10"

"d10=d23/2-p14/2-p1-p2/2"
"d29=d23/2-p14/2-d21-4u-p1-p2/2"
"d30=d23/2-p14/2"


"DELTA=d0*2+larger(p14,p22)-p14"
"DELTA1=d23-d21"
"DELTA2=d22-p14/2-p4-4u"
"DELTA3=d26-p16-d16-p11-12u"


aqseq 321


1 d11 ze
  d11 pl16:f3
2 d11 do:f3
3 d1 pl1:f1    fq=cnst21:f2
  2u pl2:f2
  p1 ph1
  d26 pl3:f3
  (center (p2 ph1) (p22 ph1):f3 )
  d26

  (p5 ph1):f1
  4u
```

Appendix

```
(p1 ph2):f1

4u

(p21 ph1):f3
d21
DELTA1
4u
(center (p4 ph1):f2 (p22 ph1):f3 )
d23 p14:f2

(p21 ph2):f3

(p3 ph3):f2

4u
(p14:sp5 ph1):f2
DELTA2
4u p12:f2
(p4 ph1):f2
4u
(p14:sp5 ph1):f2

DELTA2
4u p14:f2
(p3 ph2):f2

4u
30u fq=cnst22:f2

(p3 ph4):f2

d0
(center (p14:sp7 ph1):f2 (p22 ph1):f3 )
d0

4u p12:f2
(p4 ph1):f2
DELTA

(p14:sp7 ph1):f2
4u
4u p14:f2
(p3 ph1):f2

4u
30u fq=cnst21:f2
(p3 ph2):f2


DELTA2

(p14:sp5 ph1):f2
4u
4u p12:f2

(p4 ph1):f2
DELTA2

(p14:sp5 ph1):f2
4u
```

Appendix

```
  4u p14:f2

  (p3 ph1):f2

  (p21 ph5):f3
  d30
  (p14:sp5 ph1):f2
  (p1 ph2):f1
  (p2 ph1):f1
  (p1 ph2):f1
  d30
  4u p12:f2
  (center (p4 ph1):f2   (p22 ph8):f3)
  d10
  (p14:sp5 ph1):f2
  (p1 ph2):f1
  (p2 ph1):f1
  (p1 ph2):f1
  d29

  d21
  (p21 ph1):f3
  4u

  (p1 ph1)
  d26
  (center (p2 ph1)  (p22 ph1):f3 )
  d26 pl16:f3

  go=2 ph31 cpd3:f3
  d11 do:f3 mc #0 to 2
     F1PH(rd10 & rd29 & rd30 & ip4, id0)
     F2PH(ip5, id10 & id29 & dd30)
exit

ph1=0
ph2=1
ph3=0 0 0 0 0 0 0 0 2 2 2 2 2 2 2 2
ph4=0 2
ph5=1 1 3 3
ph6=2
ph7=3
ph8=0 0 0 0 2 2 2 2
ph31=0 2 2 0 0 2 2 0 2 0 0 2 2 0 0 2


;pl0 : 120dB
;pl1 : f1 channel - power level for pulse (default)
;pl3 : f3 channel - power level for pulse (default)
;pl16: f3 channel - power level for CPD/BB decoupling
;pl19: f1 channel - power level for CPD/BB decoupling
;sp5: f2 channel - shaped pulse 180 degree   (Ca off resonance)
;sp7: f2 channel - shaped pulse 180 degree   (C=O off resonance)
;p1 : f1 channel -  90 degree high power pulse
;p5 : f1 channel -  water purge pulse [1 ms]
;p2 : f1 channel - 180 degree high power pulse
;p13: f2 channel -  90 degree shaped pulse
;p14: f2 channel - 180 degree shaped pulse
;p16: homospoil/gradient pulse                            [1 msec]
;p21: f3 channel -  90 degree high power pulse
```

Appendix

```
;p22: f3 channel - 180 degree high power pulse
;p26: f1 channel -  90 degree pulse at pl19
;d0 : incremented delay (F1 in 3D)                      [3 usec]
;d1 : relaxation delay; 1-5 * T1
;d10: incremented delay (F2 in 3D) = d23/2-p14/2
;d11: delay for disk I/O                                [30 msec]
;d13: short delay                                       [4 usec]
;d21: 1/(2J(NH)                                         [5.5 msec]
;d22: 1/(4J(COCa)                                       [4 msec]
;d23: 1/(4J(NCO)                                        [12 msec]
;d26: 1/(4J'(NH)                                        [2.3 msec]
;d29: incremented delay (F2 in 3D) = d23/2-p14/2-p26-d21-4u
;d30: decremented delay (F2 in 3D) = d23/2-p14/2
;cnst21: frequency of relay nucleus
;cnst22: frequency of detected nucleus (O)
;o2p: Calpha chemical shift (cnst22)
;in0: 1/(2 * SW(Ca)) = DW(Ca)
;nd0: 2
;in10: 1/(4 * SW(N)) = (1/2) DW(N)
;nd10: 4
;in29: = in10
;in30: = in10
;NS: 8 * n
;DS: >= 16
;td1: number of experiments in F1
;td2: number of experiments in F2     td2 max = 2 * d30 / in30
;FnMODE: States-TPPI (or TPPI) in F1
;FnMODE: States-TPPI (or TPPI) in F2
;cpds1: decoupling according to sequence defined by cpdprg1
;cpd3: decoupling according to sequence defined by cpdprg3
;pcpd1: f1 channel - 90 degree pulse for decoupling sequence
;pcpd3: f3 channel - 90 degree pulse for decoupling sequence
```

Full side chain correlation experiment using H/N-INEPTs

```
;Cx(Ca)(H)NH

;p4 13C 180° pulse
;pl4 13C p90 power       during tobsy
;pl2 1H p180 power
;p5 13C 90° pulse
;p2 1H 180° pulse
;l31 MAS frequency (Hz)
;p18 13C hard pulse
;p10: water purge pulses (50ms each)
;pl10: f1 channel - power level water suppression (10kHz)
;pl21: f2 channel - power long range CP
;pl20: f1 channel - power long range CP
;p15: CP duration
```

Appendix

```
;sp10: tanhtan shape
;spf1: ramp CP
;cnst3: 13C middle of CO and Ca
;p8:  f2 channel 13C broadband 180 pulse




#include <prp15.prot>
#include <Avancesolids.incl>
#include <Delay.incl>

"d0=3u"
"d10=3u"
"d26=2.3m"
"d11=30m"
"DELTA=d0*2+larger(p14,p22)-p14"


1 ze
2 1m do:f3
    1m do:f2
  d1
  10u reset:f2 reset:f1
  10u fq=cnst1:f2
  10u pl4:f2
  10u pl19:f1
 1u  pl3:f3
  2u
  p3:f2 ph4
  d0
  (center (p14:sp5 ph1):f2 (p22 ph8):f3 )
  d0
  4u pl2:f2
  (p4 ph1):f2
  DELTA
  (p14:sp5 ph1):f2
  4u
 1u pl4:f2

   p3:f2 ph1
   0.5u pl5:f2

3    ; tobsy mixing

     (p6:spf10 ph10):f2
     (p6:spf10 ph11):f2
     (p6:spf10 ph12):f2
     (p6:spf10 ph13):f2
     (p6:spf10 ph14):f2
     (p6:spf10 ph10):f2
     (p6:spf10 ph11):f2
     (p6:spf10 ph12):f2
     (p6:spf10 ph13):f2
     (p6:spf10 ph14):f2
     (p6:spf10 ph16):f2
     (p6:spf10 ph17):f2
     (p6:spf10 ph18):f2
     (p6:spf10 ph19):f2
     (p6:spf10 ph20):f2
```

Appendix

```
    (p6:spf10 ph16):f2
    (p6:spf10 ph17):f2
    (p6:spf10 ph18):f2
    (p6:spf10 ph19):f2
    (p6:spf10 ph20):f2
lo to 3 times l1


4u pl10:f1
(p10 ph2):f1
(p10 ph1):f1
(p10 ph2):f1
(p10 ph1):f1
4u

  1u pl4:f2
p3:f2 ph2

  10u fq=cnst2:f2
  10u      pl21:f2
  10u      pl20 :f1
  2u
  (p17:spf1 ph3):f1 (p17 ph1):f2
  4u         pl1:f1
  d26        p18:f2
  (center (p2 ph1):f1 (p22 ph8):f3)
  d26           fq=cnst3:f2
  (center (p1 ph1):f1 (p21 ph5):f3)
  d10
   (center (p2 ph1):f1 (p8 ph1):f2)
  d10
  (center (p1 ph2):f1 (p21 ph1):f3)
   d26
  (center (p2 ph1):f1 (p22 ph8):f3)
   d26


  2u      pl16:f3


  0.5u cpd3:f3


  5u
  go=2 ph31
 do:f2              ;decoupler off
   1m do:f3


d11 do:f3 mc #0 to 2
    F1PH(rd10 & ip4, id0)
    F2PH( ip5, id10 & id29 & dd30 & dd31)

HaltAcqu, 1m          ;jump address for protection files
exit                  ;quit

ph1=0
ph2=1
ph3=0 0 0 0  2 2 2 2
ph4=0 2
ph5=0 0 2 2
```

Appendix

```
ph8=0 0 0 0  0 0 0 0  2 2 2 2  2 2 2 2


ph10= (360) 0
ph11= (360) 240
ph12= (360) 240
ph13= (360) 60
ph14= (360) 0
ph16= (360) 180
ph17= (360) 60
ph18= (360) 60
ph19= (360) 240
ph20= (360) 180


ph31= 0 2 2 0 2 0 0 2  0 2 2 0 2 0 0 2
```

CaCOH experiment for side chain amide assignments

```
;rl_cacoHtest2
;CaCOH
;3D sequence with
;     inverse correlation for triple resonance using inept and CP transfer
;                       steps
;
;      F2(Ca, t1) -> F2(CO, t2) -> F1(H,t3)
;
;on/off resonance Ca and C=O pulses using soft/shaped pulses
;using constant time in t1 and t2
;phase sensitive (t1)
;phase sensitive (t2)
;water suppression using water purge pulses as in Zhou and Riestra 2008


prosol relations=<triple>


#include <Avance.incl>
#include <Delay.incl>



"d11=30m"

"d13=4u"

"d23=12m"
"d26=2.3m"
"d22=1.5m"

"in29=in10"
"in30=in10"
```

Appendix

```
"in31=in10"
"in19=in0"
"in20=in0"
"in21=in0"

"d0=d22-p22/2"
"d19=d22-p22/2"
"d20=d22-p22/2"
"d21=d22-p22/2"
"d10=d22-p22/2"
"d29=d22-p22/2"
"d30=d22-p22/2"
"d31=d22-p22/2"


"DELTA=d0*2+larger(p14,p22)-p14"
"DELTA1=d23-d21"
"DELTA2=d26-p16-d16-p11-12u"


aqseq 321


1 d11 ze
  d11 pl16:f3
2 d11 do:f3
3 d1    pl3:f3

30u fq=cnst1:f2
2u pl20 :f1
2u pl4:f2
(p3 ph4):f2
d20
(p22 ph8):f3
d21
4u  pl6:f2
(p6 ph1):f2
d0
 (p22 ph1):f3
d19    pl4:f2
4u
(p3 ph2):f2
30u fq=cnst2:f2

(p3 ph5):f2
d30
(p22 ph8):f3
d31
4u pl6:f2
(p6 ph1):f2
d10
(p22 ph1):f3
d29 pl21:f2
4u

1u pl4:f2
   p3:f2 ph2

4u pl10:f1
(p10 ph2):f1
(p10 ph1):f1
```

Appendix

```
  (p10 ph2):f1
  (p10 ph1):f1
  4u

  1u p14:f2
    p3:f2 ph2

   2u pl21 :f2
   2u pl20 :f1

   (p17:spf1 ph3):f1 (p17 ph1):f2

    2u      pl16:f3
    4u
    go=2 ph31 cpd3:f3
    d11 do:f3 mc #0 to 2
        F1PH(rd10 & rd29 & rd30 & rd31 & ip4, id0 & id19 & dd20 & dd21)
        F2PH( ip5, id10 & id29 & dd30 & dd31)
  exit


  ph1=0
  ph2=1
  ph3=0 0 0 0 2 2 2 2
  ph4=0 2
  ph5=0 0 2 2
  ph8=0 0 0 0 0 0 0 0 2 2 2 2 2 2 2 2
  ph31=0 2 2 0  2 0 0 2  0 2 2 0  2 0 0 2

  ;p10 : 120dB
  ;pl1 : f1 channel - power level for pulse (default)
  ;pl2 : f2 channel - 180 soft rectang. C
  ;pl3 : f3 channel - power level for pulse (default)
  ;pl4 : f2 channel -  90 soft rectang. C
  ;pl6 : f2 channel - 180 hard rectang. C
  ;pl16: f3 channel - power level for CPD/BB decoupling
  ;pl19: f1 channel - power level for CPD/BB decoupling
  ;sp5: f2 channel - shaped pulse 180 degree  (Ca off resonance)
  ;p1 : f1 channel -  90 degree high power pulse
  ;p2 : f1 channel - 180 degree high power pulse
  ;p10: water purge pulses (50ms each)
  ;pl10: f1 channel - power level water suppression (10kHz)
  ;p3: f2 channel -  90 degree soft pulse
  ;p6: f2 channel - 180 degree hard pulse
  ;p21: f3 channel -  90 degree high power pulse
  ;p22: f3 channel - 180 degree high power pulse
  ;d0 : incremented delay (F1 in 3D)                       [3 usec]
  ;d1 : relaxation delay; 1-5 * T1
  ;d10: incremented delay (F2 in 3D) =  d23/2-p14/2
  ;d11: delay for disk I/O                                 [30 msec]
  ;d22: 1/(8J(CaCO)                                        [1.5 msec]
  ;cnst21: CO chemical shift (offset, in ppm)
  ;cnst22: Calpha chemical shift (offset, in ppm)
  ;o2p: CO chemical shift (cnst21)
  ;in0: 1/(4 * SW(Ca)) = (1/2) DW(Ca)
  ;nd0: 4
  ;in10: 1/(4 * SW(CO)) = (1/2) DW(CO)
  ;nd10: 4
  ;NS: 8 * n
  ;DS: >= 16
  ;td1: number of experiments in F1        td1 max = 2 * d19 / in19
```

Appendix

```
;td2: number of experiments in F2      td2 max = 2 * d19 / in19
;FnMODE: States-TPPI (or TPPI) in F1
;FnMODE: States-TPPI (or TPPI) in F2
;cpds1: decoupling according to sequence defined by cpdprg1
;cpd3: decoupling according to sequence defined by cpdprg3
;pcpd1: f1 channel - 90 degree pulse for decoupling sequence
;pcpd3: f3 channel - 90 degree pulse for decoupling sequence
```

RFDR-HSQC experiment, CP based

```
;rl_hsqcnoesy2inv(CP)
;3D NOESY-HSQC scheme with RFDR mixing using CP magnetisation transfer only
; H(H)NH


#include <Avance.incl>
#include <Delay.incl>


"p2=p1*2"

"d0=3u"

"d10=3u"

"d11=30m"

"d12=20u"

"d8=(1/cnst1-p3/1000000)/2"

;"cnst2=l1*(d8+p3)"

"cnst0=0"


aqseq 321


1 ze
  d11 pl16:f3
  d11 pl4:f2
2 d1 do:f3
3 d12 pl17:f3
    4u pl1:f1

  p1 ph7
  d0
  (center (p4 ph0):f2 (p21 ph8):f3 )
  d0
```

172

Appendix

```
  (p1 ph2):f1
  4u pl8:f1

4 d8
  (p3 ph11):f1
  d8  ipp11
  lo to 4 times l1

  4u pl1:f1
  (p1 ph4)

4u pl17:f3
4u pl15:f1
(p15:spf1 ph6):f3 (p15 ph0):f1

4u pl19:f1
  4u cpds1:f1 ph2

  d10

  (p4 ph0):f2

  d10 pl3:f3

  (p21 ph1):f3

  10u    do:f1
  2u fq=cnst0:f1
4u pl10:f1
(p10 ph0):f1
(p10 ph1):f1
(p10 ph0):f1
(p10 ph1):f1
4u
(p21 ph5):f3

  4u pl17:f3
  4u pl15:f1
  (p16:spf2 ph1):f3  (p16 ph1):f1
  0.4u      pl16:f3


go=2 ph31 cpd3:f3
  d1 do:f3 mc #0 to 2
     F1PH(rd10 & ip7, id0)
     F2PH(id10, ip6)
exit


ph0=0
ph1=1
ph2=0
ph3=2
ph4=1 1 1 1 3 3 3 3
ph5=0 0 2 2
ph6=0
ph7=0 2
ph8=0 0 0 0  0 0 0 0  2 2 2 2  2 2 2 2
ph31=0 2 2 0  2 0 0 2  0 2 2 0 2 0 0 2
ph11= 0 1 0 1 1 0 1 0
```

Appendix

```
;pl1 : f1 channel - power level for pulse (1H)
;pl3 : f3 channel - power level for pulse (15N)
;pl2 : f3 channel - power level for pulse (13C)
;pl16: f3 channel - power level for CPD/BB decoupling
;pl19: f1 channel - power level for CPD/BB decoupling
;pl4 : f2 channel - power level 13C
;pl8 : f1 channel - power level mixing 180 1H pulse
;p1 : f1 channel -  90 degree high power pulse
;p2 : f1 channel - 180 degree high power pulse
;p3 : f1 channel - 180 degree low power during mixing
;p4: f2 channel - 180 degree 13C pulse for inversion
;p21: f3 channel -  90 degree high power pulse
;p22: f3 channel - 180 degree high power pulse
;d0 : incremented delay (F1 in 3D)                 [3 usec]
;d1 : relaxation delay; 1-5 * T1
;d8 : half rotor period (mixing delay)
;d10: incremented delay (F2 in 3D)                 [3 usec]
;d11: delay for disk I/O                           [30 msec]
;d12: delay for power switching                    [20 usec]

;cnst1: spinning fr in Hz
;cnst2: mixing time
;in0: 1/(2 * SW(H)) = DW(H)
;nd0: 2
;in10: 1/(2 * SW(X)) = DW(X)
;nd10: 2
;NS: 8 * n
;DS: >= 16
;td1: number of experiments in F1
;td2: number of experiments in F2
;FnMODE: States-TPPI (or TPPI) in F1
;FnMODE: States-TPPI (or TPPI) in F2
;cpd3: decoupling according to sequence defined by cpdprg3
;pcpd3: f3 channel - 90 degree pulse for decoupling sequence
;cpds1: decoupling according to sequence defined by cpdprg1
;pcpd1: f1 channel - 90 degree pulse for decoupling sequence
```

HSQC-RFDR-HSQC experiment, CP based

```
;rl_hsqcnoesy(CP)
;3D HSQC-NOESY-HSQC scheme with RFDR mixing using CP magnetisation transfer
only
; (H)N(H)..(H)NH


#include <Avance.incl>
#include <Delay.incl>


"p2=p1*2"

"d0=3u"
```

Appendix

```
"d10=3u"

"d11=30m"

"d12=20u"

"d8=(1/cnst1-p3/1000000)/2"

;"cnst2=l1*(d8+p3)"

"cnst0=0"

aqseq 321


1 ze
  d11 pl16:f3
  d11 pl2:f2
2 d1 do:f3
3 d12 pl17:f3
    4u pl1:f1

  p1 ph7
 0.4u pl5:f1

  (p15:spf1 ph3):f3 (p15 ph1):f1

4u pl19:f1
   4u cpds1:f1 ph2

   d0
   (p4 ph0):f2
   d0 pl3:f3
  10u    do:f1

   4u pl17:f3
   4u pl5:f1
   (p16:spf2 ph2):f3   (p16 ph5):f1

   4u pl1:f1
   (p1 ph1):f1
   4u pl8:f1

;rfdr-mixing
4 d8
   (p3 ph11):f1
   d8   ipp11
   lo to 4 times l1

   4u pl1:f1
   (p1 ph4)

4u pl17:f3
4u pl5:f1
(p15:spf1 ph6):f3 (p15 ph0):f1

4u pl19:f1
   4u cpds1:f1 ph2
```

Appendix

```
  d10

 (p4 ph0):f2

  d10 pl13:f3

(p21 ph1):f3

  10u    do:f1
  2u fq=cnst0:f1
4u pl10:f1
(p10 ph0):f1
(p10 ph1):f1
(p10 ph0):f1
(p10 ph1):f1
4u
(p21 ph8):f3

  4u pl17:f3
  4u pl5:f1
  (p16:spf2 ph1):f3   (p16 ph0):f1
  2u         pl16:f3

go=2 ph31 cpd3:f3
  d1 do:f3 mc #0 to 2
     F1PH(rd10 & ip2, id0)
     F2PH(id10, ip6)
exit


ph0=0
ph1=1
ph2=0
ph3=2
ph4=1 1 1 1 3 3 3 3
ph5=0 0 2 2
ph6=0
ph7=0 2
ph8=0 0 0 0  0 0 0 0  2 2 2 2  2 2 2 2
ph31=0 2 2 0  2 0 0 2  2 0 0 2 0 2 2 0
ph11= 0 1 0 1 1 0 1 0


;pl1 : f1 channel - power level for pulse (1H)
;pl3 : f3 channel - power level for pulse (15N)
;pl2 : f3 channel - power level for pulse (13C)
;pl16: f3 channel - power level for CPD/BB decoupling
;pl19: f1 channel - power level for CPD/BB decoupling
;pl18: f1 channel - power level mixing 180 1H pulse
;p1  : f1 channel -  90 degree high power pulse
;p2  : f1 channel - 180 degree high power pulse
;p3  : f1 channel - 180 degree low power during mixing
;p4: f2 channel - 180 degree 13C pulse for inversion
;p21: f3 channel -  90 degree high power pulse
;p22: f3 channel - 180 degree high power pulse
;d0 : incremented delay (F1 in 3D)
;d1 : relaxation delay; 1-5 * T1
;d8 : half rotor period (mixing delay)
;d10: incremented delay (F2 in 3D)         [3 usec]
;d11: delay for disk I/O                   [30 msec]
;d12: delay for power switching            [20 usec]
```

Appendix

```
;cnst1: spinning fr in Hz
;cnst2: mixing time
;in0: 1/(2 * SW(N)) = DW(N)
;nd0: 2
;in10: 1/(2 * SW(N)) = DW(N)
;nd10: 2
;NS: 8 * n
;DS: >= 16
;td1: number of experiments in F1
;td2: number of experiments in F2
;FnMODE: States-TPPI (or TPPI) in F1
;FnMODE: States-TPPI (or TPPI) in F2
;cpds1: decoupling according to sequence defined by cpdprg1
;pcpd1: f3 channel - 90 degree pulse for decoupling sequence
;cpd3: decoupling according to sequence defined by cpdprg3
;pcpd3: f3 channel - 90 degree pulse for decoupling sequence
```

HSQC-RFDR-HSQC experiment, INEPT based

```
;rl_hsqcnoesyhsqc
;3D HSQC-NOESY-HSQC scheme with RFDR mixing using INEPT magnetisation transfer
only
;  (H)N(H)..(H)NH

;3D sequence with
;   homonuclear correlation via dipolar coupling
;   H-N correlation via double inept transfer
;phase sensitive (t1)
;phase sensitive (t2)
;using trim pulses in inept transfer
;with decoupling during acquisition
;
;A.L. Davis, J. Keeler, E.D. Laue & D. Moskau, J. Magn. Reson. 98,
;     207-216 (1992)


#include <Avance.incl>
#include <Delay.incl>


"p2=p1*2"

"p22=p21*2"

"d0=3u"
```

Appendix

```
"d10=3u"

"d11=30m"

"d12=20u"

"d26=2.7m"

"d8=(1/cnst1-p3/1000000)/2"

;"cnst2=l1*(d8+p3)"


aqseq 321

1 ze
  d11 pl16:f3
  d11 pl2:f2
2 d1 do:f3
3 d12 pl3:f3

  (p1 ph7)
  d26
  (center (p2 ph0) (p22 ph0):f3 )
  d26

 (center (p1 ph4) (p21 ph0):f3 )

  d0

   (center (p2 ph0) (p4 ph0):f2 )

  d0

  (center (p1 ph0) (p21 ph2):f3 )
  d26
  (center (p2 ph0) (p22 ph0):f3 )
  d26

  (p1 ph4):f1
  4u pl8:f1
; rfdr mixing
4 d8
  (p3 ph11):f1
  d8  ipp11
  lo to 4 times l1

  4u pl1:f1

  (p1 ph5)
  d26
  (center (p2 ph0) (p22 ph0):f3 )
  d26
    (pl1 ph5):f1
 1u
  (p1 ph4) (p21 ph0):f3

  d10
```

Appendix

```
  (center (p2 ph0) (p4 ph0):f2)

  d10

  (center (p1 ph1) (p21 ph6):f3 )
  d26
  (center (p2 ph8) (p22 ph8):f3 )
  d26
  1u pl16:f3

go=2 ph31 cpd3:f3
  d1 do:f3   mc #0 to 2
     F1PH(rd10 & ip2, id0)
     F2PH(id10, ip6)
exit


ph0=0
ph1=0 0 0 0 2 2 2 2
ph2=0
ph3=2
ph4=1
ph5=0 0 2 2
ph6=0
ph7=0 2
ph8=0 0 0 0  0 0 0 0  2 2 2 2  2 2 2 2
ph10=0
ph31=0 2 2 0  2 0 0 2  0 2 2 0  2 0 0 2
ph11= 0 1 0 1 1 0 1 0


;p11 : f1 channel - power level for pulse (default)
;pl3 : f3 channel - power level for pulse (default)
;pl16: f3 channel - power level for CPD/BB decoupling
;pl2: f2 channel - power level 13C
;pl8: f1 channel - power level mixing 180 1H pulse
;p1 : f1 channel -  90 degree high power pulse
;p2 : f1 channel - 180 degree high power pulse
;p3 : f1 channel - 180 degree low power during mixing
;p4: f2 channel - 180 degree 13C pulse for inversion
;p21: f3 channel -  90 degree high power pulse
;p22: f3 channel - 180 degree high power pulse
;d0 : incremented delay (F1 in 3D)                [3 usec]
;d1 : relaxation delay; 1-5 * T1
;d8 : half rotor period (mixing delay)
;d10: incremented delay (F2 in 3D)                [3 usec]
;d11: delay for disk I/O                          [30 msec]
;d12: delay for power switching                   [20 usec]

;d26: 1/(4J)NH
;cnst1: spinning fr in Hz
;cnst2: mixing time
;in0: 1/(2 * SW(H)) = DW(H)
;nd0: 2
;in10: 1/(2 * SW(X)) = DW(X)
;nd10: 2
;NS: 8 * n
;DS: >= 16
;td1: number of experiments in F1
;td2: number of experiments in F2
```

Appendix

```
;FnMODE: States-TPPI (or TPPI) in F1
;FnMODE: States-TPPI (or TPPI) in F2
;cpd3: decoupling according to sequence defined by cpdprg3
;pcpd3: f3 channel - 90 degree pulse for decoupling sequence
```

ACKNOWLEDGEMENTS

The outcome of this PhD thesis goes back to innumerable people that are involved in the work in one way or another. The achieved results have been made possible only through their good will and many fruitful interaction leading to the positive and supportive basis generated by the different members of the FMP and other related persons. In this section, I would like to kindly express my gratitude to these people.

I am very grateful to my supervisor Prof. Dr. Bernd Reif, who provided a remarkable amount of energy and enthusiasm to the manifold scientific questions and subjects and an outstanding commitment to the work of his students. I am particularly grateful for him being concerned about difficulties and spending a lot of his time to give advise and suggestions whenever a problem occurred. I would also like to acknowledge the freedom to join conferences and courses without restriction and to dedicate myself to whatever I considered interesting and valuable.

I am thankful for the advice and the help that I got from Dr. Barth van Rossum, especially in respect to the hardware in the spectrometer buildings. This also applies to a good mood and to always spreading a positive atmosphere. A similar gratitude is directed towards Dr. Peter Schmieder, who regularly dedicated the "5 min" I asked him for to thoroughly enlighten a scientific problem with a very competent knowledge.

I would like to thank Prof. Dr. Hartmut Oschkinat for the possibility to use state-of-the-art NMR equipment and his enthusiasm for proton detection in the solid state. Also I would like to say thank you to Uwe, Anne, and Kristina for the support with different proteins. I owe a special gratitude to Murali and Matthias, who prepared all of the $A\beta^{1-40}$ and OmpG samples.

Having a great ideological value for the realization of my scientific ambitions, Prof. Dr. Christian Griesinger was the person that gave me the perspectives of a fascinating field combining many aspects of different natural sciences.

I would like to express my gratitude to my lab mates Tomas, Andy, and Sam for the spirit at work, many laughs, the help that I received with computer programs that were new to me, and for the exchange of many helpful tips and tricks in daily PhD student life. This also applies to Dr. Veniamin Chevelkov, who showed me the first handling of an NMR spectrometer, and the other people in the lab, Juan-Miguel, Verena, Kerstin, Mangesh and Vipin. Further people of the institute and the MDC were responsible for a warm and humorous atmosphere during daily business and the way to and from the institute: Stefan Markovic, Umit, Sascha, Arne, and Trent, Nestor, Miriam, Susann, and many other people that I cannot explicitly mention here.

Acknowledgements

I am grateful for a support and many joyful moments that I spent in Berlin, during weekends and after work. This is due to Ingmar, Kristina, Joao, Paola, Amaya, and Rafael, e.g.. In a particular way, this was completed by the wonderful encounter of Mareike, who has given me many new perspectives, a new home, and a new view onto many things, including myself.

I am happy to always be able to count on my parents and my sister, referring not only to mental, physical, and financial support but also to always seeing each other soon, no matter how far we have to travel. The feeling of always being welcome at home and with them is a precious good. This especially includes my grandmother Ilse, who always appealed to the consideration of my health and a good balance between work and delight.

Thank you so much…!

Publications

PUBLICATIONS ARISING FROM THIS WORK

1. **R. Linser**, V. Chevelkov, A. Diehl, B. Reif, "Sensitivity Enhancement using Paramagnetic Relaxation in MAS Solid-State NMR of Perdeuterated Proteins", *J. Magn. Reson.* **189**, 209–216 (2007).
2. **R. Linser**, U. Fink, B. Reif, "Proton-Detected Scalar Coupling Based Assignment Strategies in MAS Solid-State NMR Spectroscopy Applied to Perdeuterated Proteins", *J. Magn. Reson.* **193**, 89-93 (2008).
3. A. Krushelnitsky, E. de Azevedo, **R. Linser**, B. Reif, K. Saalwaechter, D. Reichert, "Direct Observation of Millisecond to Second Motions in Proteins by Dipolar CODEX NMR", *J. Am. Chem. Soc.* **131**, 12097-12099 (2009).
4. **R. Linser**, U. Fink, B. Reif, "Probing Surface Accessibility of Proteins using Paramagnetic Relaxation in MAS solid-state NMR Spectroscopy", *J. Am. Chem. Soc.*, **131** (38), 13703-13708 (2009).
5. Ü. Akbey, S. Lange, W. T. Franks, **R. Linser**, K. Rehbein, A. Diehl, B. J. van Rossum, B. Reif, H. Oschkinat, „Optimum Levels of Exchangeable Protons in Perdeuterated Proteins for Proton Detection in MAS Solid-State NMR Spectroscopy", *J. Biomol. NMR*, **46**, 67-73 (2009).
6. **R. Linser**, U. Fink, B. Reif, "Narrow Carbonyl Linewidths of Proton-Diluted Proteins Facilitate NMR Assignments in the Solid State", accepted in *J. Biomol. NMR*.
7. V. Agarwal, **R. Linser**, U. Fink, K. Fälber, B. Reif, „Identification of hydroxyl protons and characterization of exchange behaviour and hydrogen bonding in a microcrystalline protein", *J. Am. Chem. Soc.*, **132** (9), 3187-3195 (2010).
8. V. Chevelkov, Y. Xue, **R. Linser**, N. Skrynnikov, B. Reif, „Comparison of solid-state dipolar couplings and solution relaxation data provides insight into protein backbone dynamics", *J. Am. Chem. Soc.*, **132** (14), 5015-5017 (2010).
9. **R. Linser**, U. Fink, B. Reif, „Detection of dynamic regions in biological solids enabled by spin-state selective NMR experiments", *J. Am. Chem. Soc.* **132** (26), 8891–8893 (2010).
10. **R. Linser**, M. Dasari, U. Fink, P. Schmieder, J.-M. Lopez del Amo, S. Marcovic, M. Hiller, H. Oschkinat, D. Oesterheld, B. Reif, "Proton detected solid state NMR of fibrillar and membrane proteins", accepted in *Angewandte Chemie* (2011).
11. **R. Linser**, B. Bardiaux, V. Higman, U. Fink, B. Reif, "Structure calculation from highly unambiguous amide and methyl ^{1}H-^{1}H distance restraints for a micro-crystalline protein with MAS solid state NMR", accepted in *J. Am. Chem. Soc.* (2011).
12. **R. Linser**, "Whole sidechain to backbone correlation in perdeuterated proteins through combined excitation and long-range magnetization transfer.", submitted.

CONTRIBUTIONS TO CONFERENCES

July 2007: **R. Linser**, V. Chevelkov, A. Diehl, B. Reif, "^{1}H detected ssNMR of protein mycrocrystals involving paramagnetic ions", Euromar NMR conference, Tarragona, Spain. (Poster Presentation)

September 2007: **R. Linser**, V. Chevelkov, A. Diehl, B. Reif, "^{1}H detected ssNMR of protein mycrocrystals involving paramagnetic ions", Annual meeting "Fachgruppe Magnetische Resonanz" of the German Chemical Society (GDCh), Göttingen, Germany. (Poster Presentation)

March 2008: **R. Linser**, U. Fink, B. Reif, "^{1}H-detected Scalar Coupling Based Assignment in MAS Solid-State NMR", 49th Experimental NMR Conference (ENC) 2008, Asilomar, California, USA. (Poster Presentation)

March 2009: **R. Linser**, M. Dasari, B. Reif, "Proton detection in solids – application of new assignment experiments to Alzheimer's peptide Aβ^{1-40}", 50th Experimental NMR Conference (ENC) 2009, Asilomar, California, USA. (Poster Presentation)

October 2009: J.-M. Lopez del Amo, **R. Linser**, M. Dasari, B. Reif, "Dynamic Nuclear Polarization of DNP Aβ^{1-40} – applying DNP to the Alzheimer's peptide", Safed Summer School on DNP 2009 (Dynamic Nuclear Polarization at high magnetic fields - theory and applications), Safed, Israel. (Poster Presentation)

April 2010: **R. Linser**, M. Dasari, M. Hiller, H. Oschkinat, B. Reif, "Proton detection in rotating solids – new methodology and application", 51th Experimental NMR Conference (ENC) 2010, Daytona Beach, Florida, USA. (Poster Presentation)

December 2010: **R. Linser**, B. Bardiaux, V. Higman, B. Reif, "Structure elucidation from highly unambiguous restraints of perdeuterated proteins in the solid state", Pacifichem 2010, Honolulu, Hawaii, USA. (Invited Talk)

I want morebooks!

Buy your books fast and straightforward online - at one of world's fastest growing online book stores! Environmentally sound due to Print-on-Demand technologies.

Buy your books online at
www.morebooks.shop

Kaufen Sie Ihre Bücher schnell und unkompliziert online – auf einer der am schnellsten wachsenden Buchhandelsplattformen weltweit! Dank Print-On-Demand umwelt- und ressourcenschonend produziert.

Bücher schneller online kaufen
www.morebooks.shop

KS OmniScriptum Publishing
Brivibas gatve 197
LV-1039 Riga, Latvia
Telefax: +371 686 204 55

info@omniscriptum.com
www.omniscriptum.com

Printed by Books on Demand GmbH, Norderstedt / Germany